symposia on theoretical physics 3

Contributors to this volume:

W. Brenig
A. Fujii
J. Lukierski
R. Oehme
S. K. Srinivasan
K. Symanzik
K. Venkatesan

symposia on theoretical physics

Lectures presented at the
1964 Summer School
of the Institute
of Mathematical Sciences
Madras, India

3

Edited by
ALLADI RAMAKRISHNAN
Director of the Institute

PLENUM PRESS • NEW YORK • 1967

ISBN-13: 978-1-4684-7757-3 e-ISBN-13: 978-1-4684-7755-9
DOI: 10.1007/ 978-1-4684-7755-9

Library of Congress Catalog Card Number 65-21184

Softcover reprint of the hardcover 1st edition 1976
A Division of Plenum Publishing Corporation
227 West 17 Street, New York, N. Y. 10011

Introduction

The third volume of the series entitled "Symposia on Theoretical Physics" comprises the lectures delivered at the First Matscience Summer School on Theoretical Physics held in Bangalore for three weeks from August 24 to September 13, 1964.

The academic program of the summer school consisted mainly of several invited lectures by both foreign and Indian scientists. Among the participants were the following: Professor R. Oehme, University of Chicago (United States); Professor K. Symanzik, New York University (United States); Professor E. R. Caianiello, Director, Institute for Theoretical Physics, Naples (Italy); Professor W. Brenig, Max Planck Institute (West Germany); Professor F. Calogero, University of Rome (Italy); Dr. A. Fujii, School of Science and Technology, Sophia University (Japan); Dr. J. Lukierski, University of Wroclaw (Poland). All were visiting scientists at Matscience.

It was a very fortunate circumstance that this summer school was held immediately after the International Conference on High-Energy Physics at Dubna, U.S.S.R. Among the various topics discussed at the Dubna Conference, of particular interest is the reported violation of CP invariance and, hence, the violation also of time reversal (T) invariance in some weak interactions as well as successful demonstration of SU_3 symmetry of elementary particle interactions. These were summarized by Professor Ramakrishnan who had attended the Dubna Conference.

The other main topic about which there were several lectures in the summer school was Green's function formalism, both in elementary particle physics and in many-body systems. These lectures also focused attention on some of the unsolved problems in quantum field theory.

The session began with the lectures of Caianiello, who dealt with the basic problem of formulating renormalization in a mathematically rigorous manner using algorithms like Pfaffians and Hafnians. The lectures of Symanzik dealt with the many-particle structure of Green's functions using functional formalism. He discussed the advantages of formulating the problem in terms of retarded Green's functions instead of the usual causal functions.

Oehme gave a series of twelve lectures on strong and weak interactions of elementary particles. In the former, he discussed the application of dispersion theory to high-energy scattering. He stressed that the unique interpolation of the physical partial-wave amplitudes by an analytic function of the complex angular-momentum variables has made it possible, in principle, to correlate the results of high-energy experiments with our partial knowledge of low-energy resonances and the forces in the crossed reaction. He also gave four lectures on selected topics on weak interactions, with particular reference to their underlying symmetries.

Fujii delivered a systematic series of five lectures on μ-capture in light nuclei. He discussed the present status of the experimental and theoretical aspects of μ-capture, and suggested possible modifications in the calculations.

Calogero described a novel approach to scattering theory called the "phase method," wherein differential equations are directly derived for the phase shifts of different partial waves themselves. These equations are first-order differential equations of the Riccati type. The solution of this nonlinear equation was obtained iteratively, using variational techniques.

Venkatesan, stimulated by the recent papers by Weinberg, gave two lectures on theories of particles with arbitrary spin. In the first, he carefully surveyed the various approaches, pointing out the difficulties in each of them. After presenting Weinberg's work, he also discussed the group theoretic aspects of complex angular momentum and Regge poles. Lukierski discussed various aspects of classical quantum gauge transformations and obtained the commutation relations for the Yang–Mills field.

Brenig spoke about the work of Migdal on the theory of finite nuclei using interacting quasiparticles. He also discussed the lifetime of a quasiparticle. Vasudevan reviewed Kubo's formalism, which links transport coefficients and equilibrium correlation functions of a many-body system. Ranganathan described the recent program initiated by Martin and de Dominicis, the purpose of which is to eliminate the bare potentials and formulate every physical quantity in terms of distribution functions which are the effects due to interactions.

Srinivasan gave two lectures on applications of stochastic theory to some physical problems. The first one dealt with the explanation of shot effect, introducing non-Markovian features. In his second talk, he presented his recent work with Vasudevan on Barkhausen noise in ferromagnets, for which case also the non-Markovian feature must be considered.

Besides these there were also a few other lectures and seminars delivered in the summer school which have not been included in the volume for editorial reasons.

Alladi Ramakrishnan

Contents

Contents of Other Volumes xi

Broken Symmetries, Current Algebras, and Weak Interactions. 1
Reinhard Oehme, University of Chicago, Chicago, Illinois

Partial Muon Capture in Light Nuclei 39
A. Fujii, Sophia University, Tokyo, Japan

Quantum Gauge Transformations.......................... 63
Jerzy Lukierski, University of Wroclaw, Wroclaw, Poland

Theories of Particles of Arbitrary Spins 75
K. Venkatesan, Matscience, Madras, India

Bethe–Salpeter Equation and Conservation Laws in Nuclear Physics 99
W. Brenig, Max Planck Institute for Physics, Munich Germany

On a Class of Non-Markovian Processes and Its Application to the Theory of Shot Noise and Barkhausen Noise 107
S. K. Srinivasan, Indian Institute of Technology, Madras India

Many–Particle Structure of Green's Functions............... 121
K. Symanzik, New York University, New York, New York

Author Index 171

Subject Index.. 175

Contents of Other Volumes

VOLUME 1

Symmetries and Resonances
T. K. Radha

Group Symmetries with R-Invariance
R. E. Marshak

Regge Poles and Resonances
T. K. Radha

On Regge Poles in Perturbation Theory and Weak Interactions
K. Raman

Determination of Spin-Parity of Resonances
G. Ramachandran

Pion Resonances
T. S. Santhanam

Pion–Nucleon Resonances
K. Venkatesan

The Influence of Pion–Nucleon Resonance on Elastic Scattering of Charged Pions by Deuterons
V. Devanathan

Pion–Hyperon Resonances
R. K. Umerjee

Some Remarks on Recent Experimental Data and Techniques
E. Segre

On New Resonances
B. Maglić

The Higher Resonances in the Pion–Nucleon System
G. Takeda

VOLUME 2

Origin of Internal Symmetries
E. C. G. Sudarshan

Construction of the Invariants of the Simple Lie Groups
L. O'Raifeartaigh

On Peratization Methods
N. R. Ranganathan

Large-Angle Elastic Scattering at High Energies
R. Hagedorn

Crossing Relations and Spin States
M. Jacob

The Multiperipheral Model for High-Energy Processes
K. Venkatesan

Regge Poles in Weak Interactions and in Form Factors
K. Raman

Some Applications of Separable Potentials in Elementary Particle Physics
A. N. Mitra

Form Factors of the Three-Nucleon Systems H^3 and He^3
T. K. Radha

Muon Capture by Complex Nuclei
V. Devanathan

Electrodynamics of Superconductors
B. Zumino

"Temperature Cutoff" in Quantum Field Theory and Mass Renormalization
S. P. Misra

Recent Developments in the Statistical Mechanics of Plasmas
H. DeWitt

Effective-Range Approximation Based on Regge Poles
B. M. Udgaonkar

Some Current Trends in Mathematical Research
M. H. Stone

Semigroup Methods in Mathematical Physics
A. T. Bharucha-Reid

Introduction to Quantum Statistics of Degenerate Bose Systems
F. Mohling

Recent Mathematical Developments in Cascade Theory
S. K. Srinivasan

Theory of a General Quantum System Interacting with a Linear Dissipation System
R. Vasudevan

VOLUME 4

Introductory Address
V. Weisskopf

A New Approach to Scattering Theory
R. Blankenbecler

Conserved Vector Currents and Broken Symmetries
Ph. Meyer

Multiplet Structure and Mass Sum Rules in the $SU(6)$ Symmetry Scheme
V. Singh

Group Representations for Complex Angular Momentum
K. Venkatesan

A Model of a Unitary S-Matrix for Peripheral Interactions
K. Dietz

Equivalent Potential Approach for Strong Interactions
L. A. P. Balazs

"Repulsive" Potential Approach to Pion Resonances
A. N. Mitra

The Renormalizability of Higher Spin Theories
J. Lukierski

Muon Capture by Complex Nuclei
V. Devanathan

Comments on Sum Rules
P. T. Landsberg and D. J. Morgan

Inelastic Neutron Scattering and Dynamics in Solids and Liquids
A. Sjölander

Axioms and Models
M. H. Stone

Characters of Semi-Simple Lie Groups
Harish-Chandra

Sequent Correlations in Evolutionary Stochastic Point Processes
S. K. Srinivasan

VOLUME 5

Lectures on Nested Hilbert Spaces
A. Grossmann

Elementary Homology Theory
N. R. Ranganathan

Application of Algebraic Topology to Feynman Integrals
V. L. Teplitz

Weak Interactions
R. J. Oakes

Fundamental Multiplets
A. Ramakrishnan

On the Conformal Group and its Equivalence with the Hexa-dimensional Pseudo Euclidean Group
M. Baktavatsalou

Certain Extremal Problems
K. R. Unni

Fluctuating Density Fields and Fokker–Planck Equations
S. K. Srinivasan

Contraction of Lie Groups and Lie Algebras
K. Venkatesan

The Ground States of He^3 and H^3
K. Ananthanarayanan

Remarks of the Present State of General Relativity
S. Kichenassamy

A Pattern in Functional Analysis
J. L. Kelley

Correspondence Principles in Dynamics
R. Arens

VOLUME 6

On Locally Isomorphic Groups and Cartan–Stiefel Diagrams
B. Gruber

Linear Response, Bethe–Salpeter Equation, and Transport Coefficients
L. Picman

The Description of Particles of Arbitrary Spin
P. M. Mathews

Radiative Corrections in β-Decay
G. Källén

What Are Elementary Particles Made Of?
E. C. G. Sudarshan

Recent Developments in Cosmology
J. V. Narlikar

An Introduction to Nevanlinna Theory
W. K. Hayman

Normalization of Bethe–Salpeter Wave Functions
Y. Takahashi

Non-Lagrange Theories and Generalized Conservation
Y. Takahashi

Cosmic X-Rays, γ-Rays, and Electrons
R. R. Daniel

β-Decay and μ-Capture Coupling Constants
S. C. K. Nair

Functions of Exponential Type
K. R. Unni

A Model for Processing Visual Data with Applications to Bubble Chamber Picture Recognition (Summary)
R. Narasimhan

On Functional Methods in the S-Matrix Theory
J. Rzewuski

An Impact Parameter Formalism
T. Kotani

Some Properties of the Fourier–Bessel Transform
G. Källén

The A_1 and K** (1320) Phenomena — Kinematic Enhancements of Mesons?
G. and S. Goldhaber

The Photoproduction and Scattering of Pions from H^3 and He^3
K. Ananthanarayanan

Relativistic Extensions of $SU(6)$
H. Ruegg

A Survey of π-N Scattering, and of the $T=\frac{1}{2}$ Amplitudes
B. J. Moyer

Broken Symmetries, Current Algebras, and Weak Interactions*

REINHARD OEHME†

UNIVERSITY OF CHICAGO
Chicago, Illinois

1. INTRODUCTION

The phenomenological analysis of hadron–lepton interactions, based upon current algebras and higher symmetries, shows certain rather astonishing regularities. Therefore, it is tempting to consider these processes separately in the hope of finding out more about the higher symmetries of hadrons and leptons.

In this article we shall consider the semileptonic and leptonic interactions on the basis of current algebras. From the currents involved in lepton–lepton couplings we obtain the algebra of $U(4)$, provided we take all four components of the four-vector currents in Lorentz space. Six of the generators are involved in the weak interactions, and the other six are related to the $(1 + \gamma_5)$-projections of the electromagnetic and the lepton-number currents. Corresponding $U(4)$-algebras can be constructed from the hadron currents either by starting with the total current involved in semileptonic interactions or by considering strangeness-changing and nonchanging

*Based on a paper entitled "Broken Symmetries and Leptonic Weak Interactions," which appeared in the *Annals of Physics*, Vol. 33, No. 1 (June 1965). Reproduced by permission of Academic Press Inc.

†The Enrico Fermi Institute for Nuclear Studies and the Department of Physics.

currents separately. In this way, we obtain two models for universal weak interactions; one involves a parameter θ corresponding to the Cabibbo angle, which has nothing to do with weak interactions; in the other model we have no such parameter.

Our use of $U(4)$-algebras requires that the Cabibbo angle θ be the same for vector and axial-vector currents. If we express the hadron currents in terms of a triplet of basic spinor fields with charges $z + 1$, z, and z, our first model gives preference to $z = 0$ and the second one to $z = -1$.

In Section 5 a brief discussion of the phenomenological situation concerning leptonic decays of mesons and baryons is presented. We emphasize the specific assumptions involved in the determination of the angle θ in the Cabibbo model, or of the universal suppression factor for strangeness-changing decays in the other model. Then we consider in detail the question of renormalization effects and the possible interpretation of the damping factor as a consequence of symmetry breaking terms in the strong interactions. In this connection it is of special interest to extract some of the information which is contained in the commutation relations of the currents. In the final section of this paper we shall discuss results which can be obtained if these commutation relations are combined with the hypothesis of a partially conserved axial-vector current.

In this paper nonleptonic interactions will not be considered. It seems quite plausible that these interactions involve additional elements which have no manifestation in the coupling between hadrons and leptons, and, therefore, it is presumably reasonable to consider the latter by themselves.

2. LEPTON CURRENTS

The phenomenology of leptonic reactions seems to be well described by a current–current type interaction in which the lepton fields appear in the combinations[1]

$$\bar{\nu}_l \gamma_\alpha (1 + \gamma_5) l \tag{1}$$

Here we denote the field operator for electrons or muons by $l(x) = e(x)$ or $\mu(x)$, which is the sum of annihilation operators for e^- or μ^- and creation operators for e^+ or μ^+, respectively. Correspondingly, the operators $\nu_l(x) = \nu_e(x)$ or $\nu_\mu(x)$ create antineutrinos or destroy neutrinos, and we have the anticommutation relations

$$\{l(x), \bar{l}'(x')\}_{x_0=x_0'} = \gamma_4\delta(\bar{x} - \bar{x}')\delta_{ll'}$$
$$\{\nu_l(x), \bar{\nu}_{l'}(x')\}_{x_0=x_0'} = \gamma_4\delta(\bar{x} - \bar{x}')\delta_{ll'} \tag{2}$$

and all others anticommute.

It is a very interesting and mysterious experimental result that electrons and muons, in spite of the large mass difference between them, are indistinguishable as far as the form and the strength of their interactions are concerned. We express this situation by requiring that the interaction Lagrangian always contains the lepton currents in the combination

$$l_\alpha = i\bar{\nu}_e\gamma_\alpha(1 + \gamma_5)e + i\bar{\nu}_\mu\gamma_\alpha(1 + \gamma_5)\mu \tag{3}$$

For reactions involving only leptons, we then write

$$\mathscr{L}_w = \frac{-G}{\sqrt{2}} l_\alpha l_\alpha^+ \tag{4}$$

and the decay $\mu^+ \to e^+ + \nu_e + \bar{\nu}_\mu$ is described by the matrix element of the term

$$\frac{G}{\sqrt{2}}[\bar{\nu}_e\gamma_\alpha(1 + \gamma_5)e][\bar{\mu}\gamma_\alpha(1 + \gamma_5)\nu_\mu]$$

which gives $\Gamma = G^2 m_\mu^5/192\pi^3$, $G \cong 10^{-5} m_p^{-2}$. The current (3) contains only left-handed neutrinos, and the electron as well as the μ-meson fields enter in the form

$$e_L = \frac{1}{2}(1 + \gamma_5)e \qquad \mu_L = \frac{1}{2}(1 + \gamma_5)\mu$$

As far as their masses can be neglected, the particles e^- and μ^- involved in the weak interactions have negative helicities.

Let us now consider an algebra associated with the current l_α. Out of the four lepton fields we form the spinors

$$\chi_e = \begin{pmatrix}\nu_e \\ e\end{pmatrix} \qquad \text{and} \qquad \chi_\mu = \begin{pmatrix}\nu_\mu \\ \mu\end{pmatrix}$$

and define the current densities

$$l_{i\alpha}^{(+)} = \frac{i}{2}\bar{\chi}\gamma_\alpha\frac{(1 + \gamma_5)}{2}\tau_i\chi \qquad \chi = \chi_e, \chi_\mu \tag{5}$$

Then the currents are

$$L_i = -i\int d^3x\, l_{i4}^{(+)}(x) \tag{6}$$

and are the generators of an SU_2-algebra, that is they satisfy the commutation relations

$$[L_i, L_j] = i\epsilon_{ijk} L_k \tag{7}$$

like the components of an angular momentum. The electron and muon parts of the currents are assumed to commute with each other, and therefore their sum also satisfies the commutation relations (7). We can write the current l_α in the form

$$l_\alpha = 2(l_{1\alpha}^{(+)} + il_{2\alpha}^{(+)})_e + 2(l_{1\alpha}^{(+)} + il_{2\alpha}^{(+)})_\mu \tag{8}$$

There is also, however, the neutral current

$$2l_{3\alpha}^{(+)} = \frac{i}{2}[-\bar{e}\gamma_\alpha(1+\gamma_5)e + \bar{\nu}_e\gamma_\alpha(1+\gamma_5)\nu_e - \bar{\mu}\gamma_\alpha(1+\gamma_5)\mu + \bar{\nu}_\mu\gamma_\alpha(1+\gamma_5)\nu_\mu] \tag{9}$$

which does not seem to have a direct physical significance. From high-energy neutrino experiments we know that the occurrence of neutral lepton currents in the weak couplings is very unlikely.[2] This fact is a major stumbling block in attempts to construct symmetry schemes for leptons, and perhaps it is an indication that we should consider electromagnetic and weak interactions of leptons simultaneously.[3] Let us introduce another neutral current $l_\alpha^{(0)}$ by replacing in equation (5) the τ_i by the unit matrix $\tau_0 = 1$. Then we can combine $l_\alpha^{(3)}$ and $l_\alpha^{(0)}$ to give

$$l_{Q\alpha}^{(+)} = l_{3\alpha}^{(+)} - l_{0\alpha}^{(+)} = -i\bar{e}\gamma_\alpha\frac{1+\gamma_5}{2}e - i\bar{\mu}\gamma_\alpha\frac{1+\gamma_5}{2}\mu \tag{10}$$

which is the negative helicity part of the electromagnetic lepton current, just as $2l_\alpha^{(0)}$ is the corresponding part of the lepton-number current. All of our currents $l_\alpha^{(j)}$ are conserved only in the limit $m_e = m_\mu = 0$; and so we have, for example,

$$\partial_\alpha l_{Q\alpha}^{(+)} = im_e\bar{e}\gamma_5 e + im_\mu\bar{\mu}\gamma_5\mu \tag{11}$$

We may assume that the weak interactions are mediated by vector bosons. At present, the neutrino experiments indicate[2] that the mass of these W-mesons must be larger than about 2 BeV. This could mean that the weak interactions are a manifestation of the physics in rather small dimensions, which somehow exhibits a γ_5-symmetry. This symmetry is only broken at lower energies due to the lepton masses. At sufficiently high energies we can split the electromagnetic and the lepton-number current into two *separately conserved parts* involving left-handed leptons (right-handed antileptons) and right-handed leptons (left-handed antileptons), respectively. The $(1+\gamma_5)$ parts of these currents then form an algebra together with the charged currents involved in the weak interactions.

Instead of considering only the fourth components of the currents $l_{i\alpha}$, $i = 0, 1, 2, 3$, it may be of interest to study the algebra generated by the quantities

$$L_{ia}^{(+)} = \int d^3x l_{ia}^{(+)}(x) \tag{12}$$

where $i = 0, 1, 2, 3$, and $a = 0, 1, 2, 3$. These 16 operators are the space integrals of the densities

$$-\frac{1}{2}\chi^{+}\sigma_k\tau_i\frac{1+\gamma_5}{2}\chi$$

and

$$\frac{1}{2}\chi^{+}\tau_i\frac{1+\gamma_5}{2}\chi \tag{13}$$

where $i = 0 \ldots 3$, and $k = 1 \ldots 3$.

They generate the algebra of $U(4)$. Note that we have used the hermitian densities $l_{i0}^{(+)}$ instead of the antihermitian fourth components. In this way we obtain just the sixteen generators of the *compact* group $U(4)$, and the current components form a complete set in this sense. With $l_{i4}^{(+)}$, we would have to double the number of terms in order to obtain the closed algebra of $GL(4)$, which is only locally compact. Note also that the components of the vector current (or the axial-vector currents)

$$\frac{i}{2}\bar{\chi}\gamma_\alpha\tau_i\chi \qquad \left(\text{or } \frac{i}{2}\bar{\chi}\gamma_\alpha\gamma_5\tau_i\chi\right)$$

alone do not form a complete set of generators. Only if both sets are taken together do we obtain the minimal algebra* $U_L(4) \times U_R(4)$ with the currents

$$\frac{i}{2}\bar{\chi}\gamma_\alpha\frac{1\pm\gamma_5}{2}\tau_i\chi$$

and for the $1+\gamma_5$(or $1-\gamma_5$)-projection alone, we have a one-to-one correspondence between the generators of $U_L(4)$ and the components of the currents.

We have mainly considered here the current algebra of $U_L(4)$ for leptons in order to use it later in connection with similar $U_L(4)$-algebra generated by the hadron currents as sub-algebras of $U_L(6)$ or other higher symmetries.[5] We do not try to construct

* Some general aspects of minimal algebras have been discussed by Gell-Mann and Ne'eman.[4]

$U(4)$-invariant lepton-interactions; already the kinetic terms in the Lagrangian cannot be made invariant.

3. HADRON CURRENTS

In analogy to the interaction (4) between leptons, we write the semileptonic interaction in the form of a phenomenological current-current coupling

$$\mathscr{L}_w = -\frac{G}{\sqrt{2}} J_\alpha l_\alpha^+ + h.c. \tag{14}$$

where l_α is the lepton current defined in equation (3). We do not write out the hadronic weak current J_α in terms of the field operators of physical particles; rather, we characterize its parts by the quantum numbers which are conserved in strong interactions. From the existence of decays like $n \to p + e^- + \bar{\nu}_e$, $\pi^- \to \pi^0 + e^- + \bar{\nu}_e$, and $\Sigma^- \to \Lambda + e^- + \bar{\nu}_e$ we know that we must have a strangeness conserving term with $|\Delta I| = 1$. The dominant strangeness-changing decays have $|\Delta I| = 1/2$ and $\Delta S = \Delta Q$; some examples are given by

$$\begin{array}{ll} \Lambda \to p + e^- + \bar{\nu}_e & \Xi^- \to \Lambda + e^- + \bar{\nu}_e \\ \Sigma^- \to n + e^- + \bar{\nu}_e & K^+ \to \pi^0 + e^+ + \nu_e \\ \Xi^0 \to \Sigma^+ + e^- + \bar{\nu}_e & K^0 \to \pi^+ + e^+ + \nu_e \end{array}$$

On the other hand, strangeness-changing decays with $\Delta S = -\Delta Q$ seem to have considerably smaller amplitudes[6]; perhaps such terms are related to possible CP-violating interactions which are weaker than the normal weak couplings.[7] Here we ignore such contributions and consider only the currents $J_\alpha^{[0]}$ aud $J_\alpha^{[1]}$ with $\Delta I = 0$, $|\Delta I| = 1$ and $\Delta Y = \Delta Q$, $|\Delta I| = 1/2$, respectively. As in the case of the leptons, we are interested here also in the minimal algebras which can be generated by the components of these currents. Let us first take the strangeness-conserving current by itself. For its vector part, we have the well-known hypothesis that it is proportional to a component of the isospin current $j_{i\alpha}$;

$$V_\alpha^{[0]} = 2(j_{1\alpha} + ij_{2\alpha}) \tag{15}$$

where

$$[I_i, I_j] = i\epsilon_{ijk} I_k$$

with

$$I_i = -i \int d^3 x j_{i4}(x) \qquad \text{and} \qquad i, j, k = 1 \ldots 3 \tag{16}$$

In the absence of electromagnetic interactions, the isospin current is conserved and hence the generators I_i are time independent.

The axial-vector part of the current $J_\alpha^{[0]}$ is also given by

$$A_\alpha^{[0]} = 2(j_{1\alpha}^5 + i j_{2\alpha}^5) \tag{17}$$

where

$$I_i^5(x_0) = -i \int d^3 x j_{i4}^5(x)$$

transforms like an isovector, that is,

$$[I_i, I_j^5] = i\epsilon_{ijk} I_k^5 \tag{18}$$

Note that I_j^5 is time dependent because the axial current is not conserved. We can close the algebraic system by requiring for I_j^5 the commutation rules

$$[I_i^5, I_j^5] = i\epsilon_{ijk} I_k \tag{19}$$

Then, equations (15), (18), and (19) define the algebra of $U_L(2) \times U_R(2)$[8] with the two commuting angular momenta

$$I_i^{(\pm)} = \tfrac{1}{2}(I_i \pm I_i^5) \tag{20}$$

which are related by a parity transformation.[8] We then have

$$-i \int d^3 x J_4^{[0]}(x) = 2(I_1^{(+)} + i I_2^{(+)}) \tag{21}$$

or

$$J_\alpha^{[0]} = 2(J_{1\alpha} + i J_{2\alpha}) \qquad J = j + j^5 \tag{22}$$

in analogy to equation (8) for the leptons.

Although we do not want to write out the currents in terms of the field operator of physical particles, it is extremely helpful for the discussion of the algebras generated by the weak hadron currents to use models in which these currents are given explicitly in terms of certain basic spinor fields.[9,10] We need at least three such fields in order to obtain the currents $J_\alpha^{[0]}$ and $J_\alpha^{[1]}$. Let us denote these fields by φ_1, φ_2, and φ_3, and give them the charges $z + 1$, z, and z and the strangeness 0, 0, and -1. For $z = 0$ we have then the analogy to p, n, and Λ. The weak hadron currents are given by

$$\begin{aligned} J_\alpha^{[0]} &= i(\bar{\varphi}_1 \gamma_\alpha (1 + \gamma_5) \varphi_2) \\ J_\alpha^{[1]} &= i(\bar{\varphi}_1 \gamma_\alpha (1 + \gamma_5) \varphi_3) \end{aligned} \tag{23}$$

and, in analogy to the leptons, we can introduce the two doublets

$$N = \begin{pmatrix} \varphi_1 \\ \varphi_2 \end{pmatrix} \qquad \text{and} \qquad M = \begin{pmatrix} \varphi_1 \\ \varphi_3 \end{pmatrix} \tag{24}$$

so that the currents (23) are the appropriate combinations of the densities

$$\begin{aligned} i^{(\pm)}_{i\alpha} &= i\bar{N}\gamma_\alpha \frac{(1 \pm \gamma_5)}{2} \frac{\tau_i}{2} N \\ v^{(\pm)}_{i\alpha} &= i\bar{M}\gamma_\alpha \frac{(1 \pm \gamma_5)}{2} \frac{\tau_i}{2} M \end{aligned} \tag{25}$$

where $i = 0, 1 \ldots 3$. For reasons of completeness, we have also introduced in equation (25) the $(1 - \gamma_5)$-projection of the current densities. Then we obtain the corresponding currents

$$I^{(\pm)}_{i\alpha}(x_0) = \int d^3x i^{(\pm)}_{a}(x)$$

and (26)

$$V^{(\pm)}_{i\alpha}(x_0) = \int d^3x v^{(\pm)}_{a}(x)$$

with $i = 0, 1 \ldots 3$ and $a = 0, 1 \ldots 3$, where $\gamma_0 = -i\gamma_4$.

The 2×16 operators $I^{(\pm)}_{ia}$ satisfy the equal-time commutation relations

$$[I^{(\pm)}_{ia}, I^{(\pm)}_{jb}] = i(\delta_{abc}\epsilon_{ijk}I^{(\pm)}_{kc} \mp \epsilon_{abc}\delta_{ijk}I^{(\pm)}_{kc}) \tag{27}$$

where δ_{ABC} is symmetric, $\delta_{ABO} = \delta_{AB}$, and $\delta_{ABC} = 0$ unless at least one index is zero. The operators $I^{(\pm)}_{ia}$ just represent the complete set of generators for the algebra of $U_L(4) \times U_R(4)$. The commutation relations (27) have been obtained using the explicit expressions (25) for the current densities and the general relations

$$\begin{aligned} \{\varphi_i(x), \bar{\varphi}_j(x')\}_{x'_0 = x_0} &= \gamma_4\delta_{ij}\delta(\bar{x} - \bar{x}') \\ [\gamma_A\tau_i, \gamma_B\tau_j] &= \tfrac{1}{2}\{\gamma_A, \gamma_B\}[\tau_i, \tau_j] + \tfrac{1}{2}[\gamma_A, \gamma_B]\{\tau_i, \tau_j\} \end{aligned} \tag{28}$$

However, we do not have to take the model seriously and we may abstract the algebraic properties of the currents and discard the model.

In addition to the currents $i^{(+)}_{1\alpha}$ and $i^{(+)}_{2\alpha}$, which are involved in the weak current (22), we have the components $i^{(+)}_{3\alpha}$ and $i^{(+)}_{0\alpha}$,* which can be combined to give the $(1 + \gamma_5)$-projection of the electro-

* Using only the algebra of $U_L(2)$ with the generators $I^{(+)}_{i0}$, $i = 0, 1 \cdots 3$, we do not fix the scale of the component $i^{(+)}_{00}$ because it commutes with all generators $I^{(+)}_{i0}$. In our case, the commutation relations (27) fix the currents $i^{(+)}_{0\alpha}$, $\alpha = 1 \ldots 3$, and $i^{(+)}_{00}$ is then determined as the fourth component.

magnetic current within the doublet N. With our choice of charges we find

$$i_{Q\alpha}^{(+)} = i_{3\alpha}^{(+)} + (2z + 1)i_{0\alpha}^{(+)} \tag{29}$$

which is equivalent to

$$i(z + 1)\bar{\varphi}_1\gamma_\alpha \frac{1 + \gamma_5}{2}\varphi_1 + iz\bar{\varphi}_2\gamma_\alpha \frac{1 + \gamma_5}{2}\varphi_2$$

Similarly, we find that $i_{0\alpha}^{(+)}$ is proportional to the $(1 + \gamma_5)$-projection of a particle number current for the I-spin doublet.

The operators $V_{i\alpha}^{(\pm)}$ in equation (26) satisfy the same commutation relations (27) as the $I_{i\alpha}^{(\pm)}$, and they also constitute a set of generators for the algebra of $U_L(4) \times U_R(4)$. The densities $v_{1\alpha}^{(+)}$ and $v_{2\alpha}^{(+)}$ give the strangeness-changing part of the weak hadron current, and the neutral densities $v_{1\alpha}^{(+)}$ and $v_{2\alpha}^{(+)}$ are again $(1 + \gamma_5)$-projections of charge and particle number current within the M-doublet. However, the $U(4)$-algebras of I- and V-currents are not independent within the framework of our model, because the spinor φ_1 is common to both sets of current densities. By considering a model with a triplet of fundamental spinor particles, we actually define the basis for a larger group of which the groups $U_L(4)$ or $U_L(4) \times U_R(4)$, which we have considered above, are subgroups.

With the three spinor fields φ_1, φ_2, and φ_3 we can construct nine different four-vector currents if only $(1 + \gamma_5)$-projections are considered. In this way we obtain 4×9 operators which can generate the algebra of $U_L(6)$, the unitary group in six dimensions, and with the zero components of the four-vector currents alone we obtain the algebra of $U_L(3)$. Let us introduce a Hermitian basis in the space of 3×3 matrices by using the familiar $U(3)$ matrices λ_i, $i = 0, 1 \ldots 8$, which satisfy the relations[8]

$$\begin{aligned} [\lambda_i\lambda_j] &= 2if_{ijk}\lambda_k \\ \{\lambda_i\lambda_j\} &= 2d_{ijk}\lambda_k \end{aligned} \tag{30}$$

with $\lambda_0 = \sqrt{2/3}\ 1$ and $d_{0ij} = \sqrt{2/3}\ \delta_{ij}$. Then we have the current densities

$$j_{i\alpha}^{(\pm)}(x) = i\bar{\varphi}(x)\gamma_\alpha \frac{1 \pm \gamma_5}{2}\frac{\lambda_i}{2}\varphi(x) \tag{31}$$

and the 2×36 Hermitian operators

$$F_{i\alpha}^{(\pm)}(x_0) = \int d^3x j_{i\alpha}^{(\pm)}(x) \tag{32}$$

with $a = 0, 1 \ldots 3$ and $i = 0, 1 \ldots 8$, which satisfy the commutation relations for equal times:

$$[F_{ia}^{(\pm)}, F_{jb}^{(\pm)}] = i(\delta_{abc} f_{ijk} F_{k0}^{(\pm)} \mp \epsilon_{abc} d_{ijk} F_{kc}^{(\pm)}) \tag{33}$$

where the symbol δ_{abc} has been defined below equation (27). The $F_{ia}^{(\pm)}$ are the generators of the algebra of $U_L(6) \times U_R(6)$. There is a one-to-one correspondence between these generators and the components of the Lorentz covariant densities (31), which may also by written in the form

$$\mp \varphi^\dagger \sigma_a \frac{1 \pm \gamma_5}{2} \frac{\lambda_i}{2} \varphi \qquad \text{for } a = 1 \ldots 3 \tag{34}$$

and

$$\varphi^\dagger \frac{1 \pm \gamma_5}{2} \frac{\lambda_i}{2} \varphi$$

Of special interest is a $U(6)$ subgroup of $U_L(6) \times U_R(6)$ with the generators

$$\begin{aligned} F_{i0} &= F_{i0}^{(-)} + F_{i0}^{(+)} = \int d^3x \varphi^\dagger \frac{\lambda_i}{2} \varphi \\ F_{ia} &= F_{ia}^{(-)} - F_{ia}^{(+)} = \int d^3x \varphi^\dagger \sigma_a \frac{\lambda_i}{2} \varphi \end{aligned} \qquad \text{for } a = 1 \ldots 3 \tag{35}$$

For the generators of $U(6)$ the direct connection with a Lorentz vector has been lost; F_{i0} is a zero component of a vector, whereas F_{ia} is the three-vector component of an axial vector. We see therefore that the $(1 \pm \gamma_5)$-projection of the current densities are the more natural quantities in the relativistic domain, which may include the weak interactions because of their apparent short range. On the other hand, the $U(6)$ symmetry described by (35) could be of importance mainly for strong interactions at low energies. For $U(6)$ it makes sense to interpret F_{0a} for $a = 1 \ldots 3$ as the spin operator whereas F_{00}, which commutes with all other generators, is proportional to the triplet number. On the other hand, for $F_{0a}^{(\pm)}$, $a = 0, 1 \ldots 3$, it is more reasonable to talk about the $(1 \pm \gamma_5)$-projection of the four-vector current associated with the triplet number, which may be identified with baryon number if we choose $z = -1/3$.

Above, we have considered $U_L(4) \times U_R(4)$ algebras corresponding to I-spin and V-spin. These are sub-algebras of $U_L(6) \times U_R(6)$, and on the basis of the commutation relations (27) and (33) we can make the identifications:

$$I_{ia}^{(\pm)} = F_{ia}^{(\pm)} \qquad \text{for } i = 1, 2, 3$$
$$I_{0a}^{(\pm)} = \frac{1}{\sqrt{3}} F_{8a}^{(\pm)} + \sqrt{\frac{2}{3}} F_{0a}^{(\pm)} \tag{36}$$

and

$$V_{1,2a}^{(\pm)} = F_{4,5a}^{(\pm)}$$
$$V_{3a}^{(\pm)} = \frac{1}{2} F_{3a}^{(\pm)} + \frac{\sqrt{3}}{2} F_{8a}^{(\pm)} \tag{37}$$
$$V_{0a}^{(\pm)} = \frac{1}{2} F_{3a}^{(\pm)} - \frac{1}{2\sqrt{3}} F_{8a}^{(\pm)} + \frac{2}{3} F_{0a}^{(\pm)}$$

There is a third inequivalent $U_L(4) \times U_{12}(4)$ subgroup which is related to U-spin. Because it contains neutral strangeness-changing currents, it is not directly involved in the weak couplings. The generators $U_{ia}^{(\pm)}$ also satisfy the commutation relations (27), and we have the identification

$$U_{1,2a}^{(\pm)} = F_{6,7a}^{(\pm)}$$
$$U_{3a}^{(\pm)} = \frac{1}{2} F_{3a}^{(\pm)} + \frac{\sqrt{3}}{2} F_{8a}^{(\pm)} \tag{38}$$
$$U_{0a}^{(\pm)} = \frac{-1}{2} F_{3a}^{(\pm)} - \frac{1}{2\sqrt{3}} F_{8a}^{(\pm)} + \frac{2}{3} F_{0a}^{(\pm)}$$

The $(1 \pm \gamma_5)$-projections of the electromagnetic current for our triplet model are given by[14]

$$j_{Q\alpha}^{(\pm)} = j_{3\alpha}^{(\pm)} + \frac{1}{\sqrt{3}} j_{8\alpha}^{(\pm)} + \left(z + \frac{1}{3}\right)\sqrt{6}\, j_{0\alpha}^{(\pm)} \tag{39}$$

Furthermore, the corresponding hypercharge and triplet number currents may be defined as

$$j_{Y\alpha}^{(\pm)} = \frac{2}{\sqrt{3}} j_{8\alpha}^{(\pm)} + C_0\sqrt{6}\, j_{0\alpha}^{(\pm)}$$
$$j_{T\alpha}^{(\pm)} = \sqrt{6}\, j_{0\alpha}^{(\pm)} \tag{40}$$

where we can choose $C_0 = 0$ for $z = -1/3$ and $C_0 = 2/3$ for integer values of z.

We see that, in general, the electromagnetic current $j_{Q\alpha}$ is not proportional to the sum of $i_{Q\alpha}$ and $v_{Q\alpha}$. This is only the case for $z = 0$ and $z = -1$ when the φ_1-field is the only charged or the only neutral one; for $z = 0$ we have also $i_{Q\alpha} = v_{Q\alpha} = j_{Q\alpha}$. Two combinations of I- and V-spin currents, which may be of interest in connection with weak and electromagnetic couplings, are the following:

$$i_{3\alpha}^{(\pm)} + v_{3\alpha}^{(\pm)} = \frac{3}{2}\left(j_{3\alpha}^{(\pm)} + \frac{1}{\sqrt{3}} j_{8\alpha}^{(\pm)}\right)$$

and (41)

$$i_{0\alpha}^{(\pm)} + v_{0\alpha}^{(\pm)} = \frac{1}{2}\left(j_{3\alpha}^{(\pm)} + \frac{1}{\sqrt{3}} j_{8\alpha}^{(\pm)}\right) + \frac{2}{3}\sqrt{6}\, j_{0\alpha}^{(\pm)}$$

The first combination is proportional to the electromagnetic current (39) for $z = -1/3$ and the second one for $z = 1$. We come back to these features of the currents in the next section.

4. UNIVERSALITY

In the previous two sections we have explored some minimal Lie-algebras which can be generated by lepton and hadron currents known to be involved in the weak interactions. We now want to use these algebras to find a reasonable definition for a "universal" weak coupling.

The interaction Lagrangian (4) between leptons defines the constant G, and we know, experimentally, that the strength of the strangeness nonchanging part of the semileptonic interaction is of the same order of magnitude. On the other hand, the strangeness-changing semileptonic transitions seem to be damped by roughly one order of magnitude. But we do not define the interactions directly in terms of the amplitudes involving the physical particles. Rather, we use the fundamental triplet of spinor fields which gives rise to the currents discussed earlier. The physical vertices are then obtained by taking the appropriate matrix elements of these currents, and they differ from the triplet couplings by coefficients which depend upon the assumed higher symmetries of the baryons and mesons as well as upon renormalization corrections. These renormalization effects, themselves, depend upon the higher symmetries and the way they are broken in the physical world.

Let us denote by $l_{i\alpha}^{(+)}$ the sum of the corresponding electron- and muon-currents which have been defined in equation (5). The lepton current l_α is then given by

$$l_\alpha = 2(l_{1\alpha}^{(+)} + i l_{2\alpha}^{(+)}) \qquad (42)$$

and the operators $L_{i\alpha}^{(+)} = \int d^3x l_{i\alpha}^{(+)}(x)$ generate the algebra of $U_L(4)$ with the equal time commutation relations

$$[L_{ia}^{(+)}, L_{jb}^{(+)}] = i(\delta_{abc}\epsilon_{ijk}L_{kc}^{+} - \epsilon_{abc}\delta_{ijk}L_{kc}^{+}) \tag{43}$$

The symbol δ_{abc} has been defined in Section 3. If we describe the lepton–hadron coupling by equation (14), the hadron current J_α is a superposition of $J_\alpha^{[0]}$ and $J_\alpha^{[1]}$. In order to fix the scale of the hadron current relative to the lepton current, we want to require that it is associated with $U(4)$-algebra in a way which is analogous to equations (42) and (43) for the leptons.* We consider two possibilities:

1. There are separate $U(4)$-algebras for the strangeness-changing and nonchanging parts of the hadron current.
2. The $U(4)$-algebra is related to the total current.

For both cases it follows from our previous considerations that a one-to-one correspondence between the components of the four-vector currents and the generators of the $U(4)$-algebra is possible only if we choose $(1 \pm \gamma_5)$-projections, which correspond to $-G_A/G_V = \pm 1$ for the triplets involved in the currents. We choose, then, $(1 + \gamma_5)$ because it is consistent with the leptonic currents and gives results nearer to those known for the physical particles. Let us consider now the two cases mentioned above.

1. In terms of I- and V-spin current densities, we have with equations (23)

$$J_\alpha^{[0]} = 2(i_{1\alpha}^{(+)} + i_{2\alpha}^{(+)})$$

and (44)

$$J_\alpha^{[1]} = 2(v_{1\alpha}^{(+)} + v_{2\alpha}^{(+)})$$

where $i_{i\alpha}^{(+)}$ and $v_{i\alpha}^{(+)}$ are given in terms of the F-spin currents by equations corresponding to (36) and (37). Individually, the sets of operators $I_{i\alpha}^{(+)}$ and $V_{i\alpha}^{(+)}$ satisfy the commutation relations (43) and generate the algebra of $U_L(4)$. Hence, we may write the interaction Lagrangian in the form

$$\mathscr{L}_W = -\frac{G}{\sqrt{2}} 2(i_{1\alpha}^{(+)} + ii_{2\alpha}^{(+)}) \cdot 2(l_{1\alpha}^{(+)} + il_{2\alpha}^{(+)})^\dagger$$
$$-\frac{G}{\sqrt{2}} 2(v_{1\alpha}^{(+)} + iv_{2\alpha}^{(+)}) \cdot 2(l_{1\alpha}^{(+)} + il_{2l}^{(+)})^\dagger + \text{h.c.} \tag{45}$$

The universality is expressed by using the same coupling constant

---

* Lepton-baryon symmetries have been studied by many authors. For a rather extensive list of references see, for example, reference 15.

G as for the lepton–lepton interaction, which is given by

$$\mathscr{L}_W = -\frac{G}{\sqrt{2}} 2(l_{1\alpha}^{(+)} + il_{2\alpha}^{(+)}) \cdot 2(l_{1\alpha}^{(+)} + il_{2\alpha}^{(+)})^\dagger \tag{46}$$

While the $i = 1, 2$ components of the currents $i_{i\alpha}^{(+)}$ and $v_{i\alpha}^{(+)}$ are directly involved in the weak interactions, the $i = 3$ and $i = 0$ components can be related to the $(1 + \gamma_5)$-projections of the electromagnetic current and, possibly, the triplet number current. Some of these possibilities have been discussed in Section 3, within the framework of the special triplet models. We mention here only that the analogy between hadron and lepton currents is especially close for a triplet with charges $0, -1, -1$ corresponding to $z = -1$. Then we find for the "electromagnetic current density" of the hadrons,

$$j_{Q\alpha}^{(+)} = (i_{3\alpha}^{(+)} - i_{0\alpha}^{(+)}) + (v_{3\alpha}^{(+)} - v_{0\alpha}^{(+)}) \tag{47}$$

which is the perfect analogy of

$$l_{Q\alpha}^{(+)} = l_{3\alpha}^{(+)} - l_{0\alpha}^{(+)}$$

for the leptons. For $z = 0$, we also have a reasonable correspondence. For any value of z the "triplet number current" is given by

$$j_{T\alpha}^{(+)} = \tfrac{1}{2}(i_{3\alpha}^{(+)} - 3i_{0\alpha}^{(+)}) + \tfrac{1}{2}(v_{3\alpha}^{(+)} - 3v_{0\alpha}^{(+)}) \tag{48}$$

as may be seen from equations (40). Of course, equations (47) and (48) are only $(1 + \gamma_5)$-projections, and we must add the corresponding $(1 - \gamma_5)$ terms in order to obtain the physical vector current densities.

2. Within the framework of the triplet model, the operators $I_{i\alpha}^{(+)}$ and $V_{i\alpha}^{(+)}$ are not independent; they do not commute with each other, but their mutual commutation relations give rise to the remaining generators of $U_L(6)$. Therefore, we obtain a different definition of universality if we require that the total hadron current in equation (14) is of the form

$$J_\alpha = 2(g_{1\alpha}^{(+)} + ig_{2\alpha}^{(+)}) \tag{49}$$

with densities $g_{i\alpha}^{(+)}$ such that the operators

$$G_{i\alpha}^{(+)} = \int d^3x g_{i\alpha}^{(+)}(x)$$

are again the generators of $U_L(4)$ satisfying the commutation relations (43).* Since the model should be a combination of strangeness-

* This requirement may be considered as a generalization of the one introduced by Gell-Mann and others [16,7] using $U(2)$-algebras.

changing and nonchanging parts, we can make the ansatz[16,18]

$$J_\alpha = \cos\theta\, J_\alpha^{[0]} + \sin\theta\, J_\alpha^{[1]} \tag{50}$$

where $J_\alpha^{[0]}$ and $J_\alpha^{[1]}$ are given by equation (44), or by the corresponding $U_L(6)$ current densities $j_{i\alpha}^{(+)}$. The generators $G_{i\alpha}^{(+)}$ of the "G-spin" subgroups of $U_L(6)$ can be obtained from the generators of I-spin subgroups by the unitary transformation[19]

$$G_{i\alpha}^{(\pm)} = \exp(-2i\theta F_{70}^{(\pm)}) I_{i\alpha}^{(\pm)} \exp(+2i\theta F_{70}^{(\pm)}) \tag{51}$$

with $F_{70}^{(\pm)} = U_{20}^{(\pm)}$. The result is

$$\begin{aligned} G_{i\alpha}^{(\pm)} &= \cos\theta\, I_{i\alpha}^{(\pm)} + \sin\theta\, V_{i\alpha}^{(\pm)} \qquad \text{for } i = 1, 2 \\ G_{3\alpha}^{(\pm)} &= \cos^2\theta\, I_{3\alpha}^{(\pm)} + \sin^2\theta\, V_{3\alpha}^{(\pm)} - \sin\theta\cos\theta\, V_{1\alpha}^{(\pm)} \\ G_{0\alpha}^{(\pm)} &= \cos^2\theta\, I_{0\alpha}^{(\pm)} + \sin^2\theta\, V_{0\alpha}^{(\pm)} + \sin\theta\cos\theta\, V_{1\alpha}^{(\pm)} \end{aligned} \tag{52}$$

and there are corresponding relations for the densities. Using equations (35)–(37) we find

$$G_{3\alpha}^{(\pm)} + G_{0\alpha}^{(\pm)} = F_{3\alpha}^{(\pm)} + \frac{1}{\sqrt{3}} F_{8\alpha}^{(\pm)} + \frac{2}{3} F_{0\alpha}^{(\pm)} \tag{53}$$

In the place of equation (45), we have now for the lepton–hadron interaction the expression

$$\mathscr{L}_W = -\frac{G}{\sqrt{2}} 2[\cos\theta\,(j_{1\alpha}^{(+)} + ij_{2\alpha}^{(+)}) + \sin\theta\,(j_{4\alpha}^{(+)} + ij_{5\alpha}^{(+)})] \times 2(l_{1\alpha}^{(+)} + il_{2\alpha}^{(+)})^\dagger + \text{h.c.} \tag{54}$$

and the universality principle is expressed by using the same coupling constant as in equation (46). The angle θ is an unspecified parameter which can be used to adjust the relative magnitude of $\Delta S = 0$ and $\Delta S = 1$ amplitudes. The effective coupling constant for neutron β-decay is now $G\cos\theta$, and for Fermi transitions there are no renormalization effects because the $\Delta S = 0$ part of the vector current is conserved in our model. This implies that θ must be small, which gives a damping factor for $\Delta S = 1$ amplitudes. We see also that we cannot recover the interaction (45) of case 1) by setting $\theta = \pi/4$ because the extra factor $1/\sqrt{2}$ would spoil our universality principle.* The first model is obtained only if we are willing to consider the $U_L(4)$ algebras or strangeness-changing and nonchanging parts of the hadron currents separately.

A priori, the $i = 3$ and $i = 0$ components of the current density

---

* A corresponding factor $1/\sqrt{2}$ for the lepton currents seems to be plausible only in an academic model with one neutrino.[17]

$g_{i\alpha}^{(+)}$ have little relation to physical quantities. However, we see from equation (50) that the combination

$$g_{3\alpha}^{(+)} + g_{0\alpha}^{(+)}$$

is independent of θ and coincides with the $(1 + \gamma_5)$ projection of the electromagnetic current $j_{Q\alpha}^{+}$ for $z = 0$.

It is important to recognize that our requirement that there exist corresponding $U_L(4)$-subalgebras for leptonic and hadronic weak currents implies that the angle θ is the same for vector and axial-vector currents.

In the discussion of cases 1) and 2) we have given some preference to triplet models with integer charges $z = -1$ or 0, because these models give rise to the $(1 \pm \gamma_5)$ projections of the electromagnetic current in a rather natural way. The reason for this can be found in the structure of the $U_{L,R}(4)$ groups as subgroups of $U_{L,R}(6)$. The current densities $i_{0\alpha}^{(\pm)}$, $v_{0\alpha}^{(\pm)}$, or $g_{0\alpha}^{(\pm)}$ all contain an additive term $\sqrt{\frac{2}{3}}\, j_{0\alpha}^{(\pm)} = \frac{1}{3} j_{T\alpha}^{(\pm)}$. However, if we are willing to require only that the equations (47) and (49) give $j_{Q\alpha}^{(+)}$ up to an additive term $\frac{2}{3} j_{T\alpha}^{(+)}$ or $-\frac{1}{3} j_{T\alpha}^{(+)}$, respectively, then we can have the quark model with $z = -\frac{1}{3}$, which may be desirable in view of recent $SU(6)$ results for the magnetic moments of baryons.

5. PHENOMENOLOGY WITH $SU(3)$

We want to consider now some of the physical consequences of interactions like (54) and (45). In these Lagrangians, the hadron current is given as a combination of the densities $j_{i\alpha}^{(\pm)}$, which give rise to the generators for the algebra of $U_L(6) \times U_R(6)$. For leptonic decays we are interested in the matrix elements of these currents with respect to the physical baryons and mesons. These matrix elements are determined by strong-interaction effects, and, in order to obtain relations between them, we must assume certain exact and approximate symmetries for these interactions. The group $U_L(6) \times U_R(6)$, although generated by our currents, need not be a good approximate symmetry for the classification of hadrons. However, there are subgroups like $U_L(3) \times U_R(3)$, $U(6)$, $U(3)$, and various $U(2)$ groups which may be more favorable. Of the $U(2)$ groups, we know that isospin is a good quantum number for the strong interactions and the corresponding vector currents $j_{i\alpha} = j_{i\alpha}^{(+)}$

$+ j_{i\alpha}^{(-)}$, $i = 1, 2, 3$ are conserved. In the following we consider $SU(3)$ as a good approximate symmetry, and we use it to classify the baryons and mesons in the conventional way. The currents $j_{i\alpha}^{(\pm)}$ satisfy the commutation relations

$$[F_i(t), j_{j\alpha}^{(\pm)}(x)]_{t=x_0} = if_{ijk} j_{k\alpha}^{(\pm)}(x)$$

with the generators of $SU(3)$, and hence they transform according to the regular representation *8* whereas $j_{0\alpha}^{(\pm)}$ transforms like a singlet *1*. In the limit of $U(3)$ symmetry we can then express the matrix elements of these currents in terms of reduced matrix element. This is just an application of the Wigner–Eckhart theorem.

In general, if we denote the state of an octet by $|O_i\rangle$, we find simply

$$\langle O_i | j_{k\alpha}^{(\pm)} | O_j \rangle = if_{ijk} M_f + d_{ijk} M_d \tag{55}$$

There can be two reduced matrix elements because the representation *8* appears twice in the decomposition

$$8 \times 8 = 1 + 8 + 8 + 10 + \overline{10} + 27 \tag{56}$$

For other cases, like those involving the decuplet, the Clebsch–Gordan coefficients can be found in the literature.[20]

The matrix elements for the physical baryons and mesons can be easily obtained from equation (55) using the identification for wave functions given by

$$\begin{pmatrix} \frac{1}{\sqrt{2}}\Sigma^\circ + \frac{1}{\sqrt{6}}\Lambda & \Sigma^+ & p \\ \Sigma^- & -\frac{1}{\sqrt{2}}\Sigma^\circ + \frac{1}{\sqrt{6}}\Lambda & n \\ \Xi^- & \Xi^\circ & -\sqrt{\frac{2}{3}}\Lambda \end{pmatrix} \tag{57}$$

or

$$\begin{pmatrix} \frac{1}{\sqrt{2}}\pi^0 + \frac{1}{\sqrt{6}}\eta_0 & \pi^+ & K^+ \\ \pi^- & -\frac{1}{\sqrt{2}}\pi^0 + \frac{1}{\sqrt{6}}\eta_0 & K_0 \\ K^- & \bar{K}^0 & -\sqrt{\frac{2}{3}}\eta_0 \end{pmatrix} \tag{58}$$

where

$$\sum_{j=1}^{8} O_j\lambda_j = \begin{pmatrix} O_3 + \frac{1}{\sqrt{3}} O_8 & O_1 - iO_2 & O_4 - iO_5 \\ O_1 + iO_2 & -O_3 + \frac{1}{\sqrt{3}} O_8 & O_6 - iO_7 \\ O_4 + iO_5 & O_6 + iO_7 & -\frac{2}{\sqrt{3}} O_8 \end{pmatrix} \tag{59}$$

Omitting unessential indices, we find for the matrix elements of strangeness-changing and nonchanging currents

$$\begin{aligned} \langle B|j_1 + ij_2|A\rangle &= if_{AB(1+i2)} M_f + d_{AB(1+i2)} M_d \\ \langle B|j_4 + ij_5|A\rangle &= if_{AB(4+i5)} M_f + d_{AB(4+i5)} M_d \end{aligned} \tag{60}$$

where the coefficients are listed in Tables I and II for $\Delta S = 0$ and for $\Delta S = 1$, respectively. Here we have used the notation $f_{AB(1+i2)} = f_{AB1} + if_{AB2}$, etc. For the pseudoscalar mesons, we can only have f-type matrix elements in equations (60) as a consequence of the Pauli principle. The Clebsch–Gordan coefficients are obtained from Tables I and II using the identification (58).

In order to compare our two models for the weak hadron current with experiments, it is convenient to use a phenomenological form for J_α. We write, for practical purposes,*

$$J_\alpha = V_\alpha + A_\alpha \tag{61}$$

where

$$\begin{aligned} V_\alpha &= (V_{1\alpha} + iV_{2\alpha}) + (V_{4\alpha} + iV_{5\alpha}) \\ A_\alpha &= (A_{1\alpha} + iA_{2\alpha}) + (A_{4\alpha} + iA_{5\alpha}) \end{aligned} \tag{62}$$

Here the vector and axial-vector currents $V_{i\alpha}$ and $A_{i\alpha}$ are proportional to $j_{i\alpha}^{(+)} + j_{i\alpha}^{(-)}$ and $j_{i\alpha}^{(+)} - j_{i\alpha}^{(-)}$, respectively. For example, the proportionality factors in the combinations (62) are simply "one" in the first model defined by equation (45), whereas for the second model we have the factors $\cos\theta$ and $\sin\theta$ according to equation (54). Note that in this equation there appears the same angle θ for vector and axial-vector terms.

Let us consider first the measurable mesons matrix elements. We obtain information about the axial vector currents from the decays $\pi^+ \to \mu^+ + \nu_\mu$ and $K^+ \to \mu^+ + \nu_\mu$. Using Lorentz-covariance

* For the remainder of this article the symbols V and A are used to denote vector- and axial-vector currents.

we can write

$$\begin{aligned}\langle o|A_{1\alpha} - iA_{2\alpha}|\pi^+\rangle &= i\pi_\alpha B_\pi \\ \langle o|A_{4\alpha} - iA_{5\alpha}|K^+\rangle &= iK_\alpha B_K\end{aligned} \tag{63}$$

where the real invariants B_π and B_K depend upon $-\pi_\alpha^2 = m_\pi^2$ and $-K_\alpha^2 = m_K^2$, respectively. With the interaction (14), the decay rate is then given by

$$\Gamma_{(\pi\to\mu+\nu)} = \frac{G^2}{8\pi} m_\pi m_\mu^2 \left(1 - \frac{m_\mu^2}{m_\pi^2}\right)^2 B_\pi^2 \cdots \tag{64}$$

and we have the ratio

$$\frac{\Gamma_{(K\to\mu+\nu)}}{\Gamma_{(\pi\to\mu+\nu)}} = \frac{B_K^2}{B_\pi^2}\frac{m_K}{m_\pi}\left(\frac{1 - m_\mu^2/m_K^2}{1 - m_\mu^2/m_\pi^2}\right)^2 \tag{65}$$

From the known decay rates for K^+ and π^+ and the $K_{\mu 2}$ branching ratio, we find

$$\frac{B_K}{B_\pi} \approx 0.27 \tag{66}$$

with an error less than ± 0.01. We note that the invariants B_K and B_π have the dimensions of an energy. We must use some common basic mass as the scale to find the ratio (66) for the dimensionless constants.

In the limit of $SU(3)$ symmetry, we have, of course, $m_\pi = m_K$ and, consequently, $B_K/B_\pi = 1$ for the first model described in Section 3, and $B_K/B_\pi = tg\,\theta$ for the second one. In case 1) the damping of the strangeness-changing transition may be understood as a consequence of the fact that $SU(3)$-symmetry is broken, whereas, in case 2) it is described by the smallness of the angle θ which has been introduced in Section 3 for algebraic reasons.

We can obtain information about the vector current from the reactions $\pi^+ \to \pi^0 + e^+ + \nu_e$ and $K^+ \to \pi^0 + e^+ + \nu_e$.[18] The relevant form factors can be defined by the covariant expansions

$$\begin{aligned}\langle \pi^0|V_{1\alpha} - iV_{2\alpha}|\pi^+\rangle &= (\pi_\alpha + \pi_\alpha^0)\sqrt{2}\, F_{\pi\pi} \\ \langle \pi^0|V_{4\alpha} - iV_{5\alpha}|K^+\rangle &= [(K_\alpha + \pi_\alpha) + (K_\alpha - \pi_\alpha)\xi_{\pi K}]\frac{1}{\sqrt{2}} F_{\pi K}\end{aligned} \tag{67}$$

where $F_{\pi\pi}$ is a function of $(\pi - \pi^0)^2$, while $F_{\pi K}$ and $\xi_{\pi K}$ depend upon the invariant $(K - \pi)^2$. Since the current $V_{i\alpha}$ for $i = 1, 2, 3$ is proportional to the conserved isospin current, it follows that there is no term proportional to $(\pi_\alpha - \pi_\alpha^0)$ in the first matrix element of equation (67), and that $F_{\pi\pi}(-q^2)$ is proportional to the electromagnetic form factor of the pion. Note that we have used the Clebsch–Gordan coefficients of Tables I and II, together with the substitution (58), in order to define the form factors such that we obtain in the $SU(3)$-limit $F_{\pi K}/F_{\pi\pi} = 1$ for model 1) and $F_{\pi K}/F_{\pi\pi} = tg\,\theta$ for model 2). Of course, we have $\xi_{\pi K} = 0$ in this limit also, because the whole octet of vector currents $V_{i\alpha}$ is conserved.

The ratio $F_{\pi K}/F_{\pi\pi}$ can be obtained from the partial decay rate for K_{e3}^{+}. In this decay the argument of $F_{\pi K}$, i.e.,

$$-(K-\pi)^2 = m_K^2 + m_\pi^2 - 2m_K E_\pi \tag{68}$$

varies between $(m_K - m_\pi)^2$ and 0 corresponding to the possible pion energies

$$m_\pi \leq E_\pi \leq \frac{m_K^2 + m_\pi^2}{2mK} \tag{69}$$

The experiments seem to be compatible with an approximately constant form factor and a ratio

$$\frac{F_{\pi K}}{F_{\pi\pi}} \simeq 0.24(\pm 0.01) \tag{70}$$

We note that in the transition amplitude for K_{e3} decay the terms involving $\xi_{\pi K}$ can be neglected because they are proportional to the square of the electron mass.

The ratio (70) for the dimensionless vector form factors is remarkably close to the corresponding ratio (66) for the axial vector invariants. But before we discuss the possible relevance of this result, we want to consider the baryonic matrix elements.

In analyzing the leptonic decays of hyperons, we do not consider the two Lagrangians of Section 3 simultaneously, because such a procedure would involve a rather cumbersome notation. Hence, at first, we discuss the phenomenological situation only for the interaction (53) involving the Cabibbo angle. Later, we can extract the implications for the other case by a simple reinterpretation of the notation.

Let A and B be two particles of the baryon octet. Then we can write, in general,

$$\begin{aligned} \langle B|V_{k\alpha}|A\rangle &= i\bar{u}(B)\{\gamma_\alpha F_{ABk}(-q^2) + \sigma_{\alpha\beta}q_\beta G_{ABk}(-q^2)\}u(A) \\ \langle B|A_{k\alpha}|A\rangle &= i\bar{u}(B)\{\gamma_\alpha\gamma_5 B_{ABk}(-q^2) + iq_\alpha\gamma_5 D_{ABk}(-q^2)\}u(A) \end{aligned} \tag{71}$$

where $q = A - B$, and where only those invariants have been used which behave under *CP*-transformations in the same way as the terms γ_α and $\gamma_\alpha\gamma_5$, respectively (1. class currents only).[24] In the limit of $SU(3)$-symmetry, we can express all form factors in equation (71) in terms of reduced invariants. So we have

$$\begin{aligned} F_{ABK}(-q^2) &= if_{ABK}F_f(-q^2) + d_{ABK}F_d(-q^2) \\ B_{ABK}(-q^2) &= if_{ABK}G_f(-q^2) + d_{ABK}G_d(-q^2) \end{aligned} \tag{72}$$

and corresponding relations for the form factors G and D. In Tables I and II we have listed the Clebsch–Gordan coefficients as far as they are relevant for weak interactions.

The octet of vector currents in our model contains the isospin current; we have $V_{k\alpha} = i_{k\alpha}$ for $k = 1, 2, 3$. We can use this fact to express the reduced vector form factors in equation (72), etc., in terms of the electromagnetic form factors of the nucleons. Although the physical baryons have triplet number zero, we must consider in general matrix elements like

$$\langle p|V_{3\alpha} + \frac{1}{\sqrt{3}}V_{8\alpha} + \left(z + \frac{1}{3}\right)\sqrt{6}\,V_{0\alpha}|p\rangle\cdots$$

For $z = -1/3$, we find the relations

$$\begin{aligned} F_{pp3} + \frac{1}{\sqrt{3}}F_{pp8} = F_f + \frac{1}{3}F_d &= F_p \\ -\frac{2}{3}F_d &= F_n \\ -\frac{1}{3}F_d &= F_\Lambda \end{aligned} \tag{73}$$

and corresponding ones for other baryons. For other values of z, a common singlet term must be added on the left-hand side. The functions $F_p(-q^2)$, $F_n(-q^2)$ and $F_\Lambda(-q^2)$ are the electromagnetic form factors of the proton, the neutron and the Λ-hyperon. For $q^2 = 0$ we find, in particular, that

$$F_f(0) = 1 \qquad F_d(0) = 0 \tag{74}$$

The singlet term vanishes for $q^2 = 0$ because the baryons have triplet number zero if $z = 0$ or $z = -1$. As may be seen from Table I, the relation $F_d(0) = 0$ also guarantees that the matrix element $\langle\Lambda|i_{1\alpha} + ii_{2\alpha}|\Sigma^-\rangle$ vanishes in the $SU(3)$-limit at $q^2 = 0$.

The functions $G_f(-q^2)$ and $G_d(-q^2)$ can be related to the magnetic form factors of the baryons:

$$G_f + \frac{1}{3} G_d + 2\left(z + \frac{1}{3}\right) G_s = G_p$$

$$-\frac{2}{3} G_d + 2\left(z + \frac{1}{3}\right) G_s = G_n \tag{75}$$

$$-\frac{1}{3} G_d + 2\left(z + \frac{1}{3}\right) G_s = G_\Lambda$$

Here we have explicitly included the singlet "magnetic" form factor $G_s(-q^2)$, which is defined in accordance with equation (71). Of some interest is the case $z = -1/3$, in which we can use the known values for the anomalous magnetic moments of the nucleons in order to compute the ratio $G_d(0)/G_f(0)$. With

$$G_p(0) = \frac{\mu_p}{2m} = \frac{1.78}{2m} \qquad G_n(0) = \frac{\mu_n}{2m} = -\frac{1.91}{2m}$$

we find[22]

$$\frac{G_d(0)}{G_f(0)} \approx 3.5 \tag{76}$$

a value which we may compare later with d/f ratios for other vertices, and which seems to be rather large.

With equation (74), the effective coupling constant for the vector part of nuclear β-decay is given by

$$G_V = G \cos\theta$$

where G must be identified with the value of G_μ obtained from μ-decay. With electromagnetic corrections included, the experimetnal ratio G_V/G_μ is given by[25]

$$\frac{G_V}{G_\mu} = 0.980 \pm 0.002 \tag{77}$$

As we have already mentioned, the universality hypothesis and the assumption of a conserved vector current imply consequently that θ must be small. Whether one should use the discrepancy between $G_V/G_\mu = 1$ and the value (77) in order to determine θ is an open question in our opinion.[26] Later we will see that the values of θ obtained from $\Delta S = 1$ decays, under the assumption that renormalization effects are small, may not be incompatible with the angle $\theta \approx 0.20$ corresponding to (77).

The effective coupling constant for the axial vector part of the nuclear β-decay amplitude is given by

$$-G_A = G \cos\theta\, [B_f(0) + B_d(0)] \tag{78}$$

where we note that our model 2 requires the same angle θ for vector and axial vector currents. Experimentally we have

$$\frac{-G_A}{G_V} = 1.15 \pm 0.05$$

which implies

$$B_f(0) + B_d(0) \approx 1.15 \tag{79}$$

The induced pseudoscalar term $D(-q^2)$ can be computed to some extent using dispersion relations (Goldberger–Treiman relation). Later we will discuss some of these relations. Now we want to work only in the allowed approximation. Using the $SU(3)$-limit exhibited in equation (72), we can obtain the matrix elements for all leptonic hyperon decays in terms of two parameters, the angle θ and the ratio $B_d(0)/B_f(0) = d/f$. The relevant expressions are found easily using Tables I and II.

Table I

Clebsch–Gordan Coefficients for $\Delta S = 0$

A	B	$if_{AB(1+i2)}$	$d_{AB(1+i2)}$
n	p	$+1$	$+1$
Σ^-	Σ^0	$\sqrt{2}$	0
Ξ^-	Ξ^0	-1	$+1$
Σ^-	Λ	0	$\sqrt{\frac{2}{3}}$
Σ^0	Σ^+	$-\sqrt{2}$	0

Table II

Clebsch–Gordan Coefficients for $\Delta S = 1$

A	B	$if_{AB(4+i5)}$	$d_{AB(4+i5)}$
Λ^0	p	$-\sqrt{\frac{3}{2}}$	$-\frac{1}{\sqrt{6}}$
Σ^-	n	-1	$+1$
Σ^0	p	$-\frac{1}{\sqrt{2}}$	$\frac{1}{\sqrt{2}}$
Ξ^-	Λ	$\sqrt{\frac{3}{2}}$	$\frac{1}{\sqrt{6}}$
Ξ^-	Σ^0	$\frac{1}{\sqrt{2}}$	$\frac{1}{\sqrt{2}}$
Ξ^0	Σ^+	$+1$	$+1$

From the branching ratios for the leptonic transitions we find

$$\frac{d}{f} \approx 2 \pm 0.4 \qquad \text{and} \qquad \theta \approx 0.26 \pm 0.02 \tag{80}$$

as a reasonable solution, which is also compatible with the decay rate for $\Sigma^- \to \Lambda + e^- + \bar{\nu}_e$.[23] For the form of the transition amplitude for $\Lambda \to p + e^- + \bar{\nu}$ we obtain with these values V-$0.64A$, which is in reasonable agreement with present experiments giving V-$(0.8 \pm 0.3)A$. Another solution, which uses all available input data, has been given by Willis *et al.*; it leads to

$$\frac{d}{f} = 1.7 \pm 0.35 \tag{81}$$

and an angle θ which is compatible with (80). The d/f-ratio obtained from leptonic decays is quite compatible with the corresponding ratio for the strong coupling between pseudoscalar mesons and baryons. This connection can be understood on the basis of the Goldberger–Treiman relations. Neglecting possible corrections due to the branch cuts, we can write these relations in the form

$$B_{AB}^{[0]} \approx \frac{g_{AB\pi}\hat{B}_\pi}{m_A + m_B} \tag{82}$$

for $\Delta S = 0$, and

$$B_{AB}^{[1]} \approx \frac{g_{ABK}\hat{B}_K}{m_A + m_B} \tag{83}$$

for $\Delta S = 1$ transitions. Here we have used the notation

$$B_{AB}^{[0]} = B_{AB1}^{(0)} + iB_{AB2}^{(0)}$$

and

$$B_{AB}^{[1]} = B_{AB4}^{(0)} + iB_{AB5}^{(0)}$$

and $\hat{B}_\pi = B_\pi/\cos\theta$, $\hat{B}_K = B_K/\sin\theta$ for model 2), or $\hat{B}_\pi = B_\pi$, $\hat{B}_K = B_K$ for model 1). The constants $g_{AB\pi}$, g_{ABK} are the pseudoscalar coupling constants, e.g., $g_{np\pi} = \sqrt{2}\, g_{NN\pi}$, $g_{\Lambda pK} = g_{\Lambda NK}$, etc. In the $SU(3)$-limit, we express the quantities B_{AB} in terms of B_f and B_d, and we do the same for the g's. Then we find

$$\frac{B_d}{B_f} \approx \frac{g_d}{g_f} \tag{84}$$

6. RENORMALIZATION EFFECTS

From the results given in the previous sections, it follows that we can obtain a very reasonable description of the leptonic interactions by using the hadron current (50) and evaluating its matrix elements in the limit of $SU(3)$-symmetry. Within this framework, the evidence for a universal damping factor for the strangeness-changing transitions seems to be rather impressive. In our model 2) this factor is given by $tg\theta$, where θ is a mixing angle which has been introduced in order to have a $U_L(4)$-algebra which is generated by the components of the total hadron-current participating in the coupling to leptons, and which is analogous to a corresponding algebra involving the components of the lepton current. We see that in this model the damping has no direct connection with the strong interactions. For some unspecified reason the weak coupling selects a component with $\theta \approx 1/4$, which remains essentially unchanged if the symmetry breaking part of the strong interaction is switched on or off. Also in this connection, it is important to remember that the unique value for $tg\theta$ has been achieved under the assumption that renormalization corrections due to the symmetry-breaking interactions are small. Perhaps this assumption is reasonable, but it may sound a bit strange in view of the rather large mass splitting within the meson octet. If we assume that there exists a strong interaction Lagrangian which can be split into an $SU(3)$ invariant and a symmetry breaking part, we can make a formal perturbation expansion in terms of the latter part. Using either dispersion theory or group theoretical methods, one can easily show that the corrections to the matrix elements of the currents are of second or higher order in γ, where γ describes the strength of the symmetry-breaking interaction.[29,30] Again, this does not imply that the corrections are small because of the appearance of large meson mass differences in the mesonic matrix elements. These mass differences are relevant also for the baryonic matrix elements because the mesons appear as pole terms in dispersion representations of the form factors.

Let us now turn to model 1) with the hadron-lepton interaction given by equation (45). Here we have true universality in the $SU(3)$-limit.[31] There is no adjustable parameter like the Cabibbo angle θ in model 1), and, as long as the symmetry is not broken, there is no damping of the strangeness-changing transitions. The idea is that this damping comes about as a consequence of the symmetry-breaking part of the strong interaction. Since this interaction must give rise to mass ratios like $m_\pi/m_K \approx 1/4$, it could also produce a damping factor which is of the same order of magnitude, although in many other instances its effects may amount only to small perturbations. The essential assumption underlying model 1) is that the universal damping of $\Delta S = 1$ transitions is a large manifestation of the symmetry-breaker, whereas the remaining corrections are small so that we can use the Clebsch–Gordan coefficients of $SU(3)$.

For the matrix elements of the axial vector currents, it is quite natural to obtain proportionality of the mass ratio m_π/m_K and the ratio of reduced matrix elements for $\Delta S = 1$ and $\Delta S = 0$ transitions. Then we have a damping factor

$$\lambda_A \approx \frac{m_\pi}{m_K} \tag{85}$$

for strangeness-changing axial-vector amplitudes.

Let us now use the G.T. relations for the corresponding baryon matrix elements.[32] We have

$$B_{AB}^{[0]} = \frac{g_{AB\pi} B_\pi}{m_A + m_B}$$

and (86)

$$B_{AB}^{[1]} = \frac{g_{ABK} B_K}{m_A + m_B}$$

The notation used in equations (84) is essentially the same as in Section 5, except that here there are no factors $\cos\theta$ or $\sin\theta$. We assume that we can neglect the baryon mass differences and that the coupling constants g_{ABK} and $g_{AB\pi}$ can be expressed in terms of the reduced quantities g_f and g_d using the Clebsch–Gordan coefficients of $SU(3)$. Then we see from equations (86) that the axial vector form factors can be expanded correspondingly in terms of $B_{d,f}^{[0]}$ and $B_{d,f}^{]1]}$, and we find the ratios

$$\frac{B_d^{[0]}}{B_f^{[0]}} \approx \frac{B_d^{[1]}}{B_f^{[1]}} \approx \frac{g_d}{g_f} \tag{87}$$

together with

$$\frac{B_{d,f}^{[1]}}{B_{d,f}^{[0]}} \approx \frac{B_K}{B_\pi} \approx \frac{m_\pi}{m_K} \tag{88}$$

Hence, we have a "universal" axial vector damping factor $\lambda_A = m_\pi/m_K$ such that $\lambda_A \to 1$ in the limit of complete $SU(3)$-symmetry. Using the unsubtracted dispersion relations for the combination of baryon form factors appearing in the divergence of the axial current and making the usual pole approximation, we can also calculate the induced pseudoscalar terms without introducing new parameters.[33]

For the vector currents we can obtain an analogous connection between baryonic and mesonic reduced form factors, provided we are willing to make pole approximation involving vector mesons like ρ and K^*.[34] Using $SU(3)$-symmetric vector meson-baryon couplings, we find the relation

$$\frac{F_f^{[1]}}{F_f^{[0]}} \approx \frac{F_{\pi K}}{F_{\pi\pi}} \frac{g_{\rho\pi\pi}}{2g_{K^*K\pi}} \tag{89}$$

where the notation is analogous to the one employed in the axial-vector case. If we assume also that $SU(3)$ symmetry is a rough zeroth approximation for the coupling between vector and pseudoscalar mesons, then we have $g_{\rho\pi\pi} \sim 2g_{K^*K\pi}$, and hence we find again a general damping factor

$$\lambda_V \approx \frac{F_{\pi K}}{F_{\pi\pi}}$$

for the vector part of the strangeness-changing decays. Although we know empirically that $\lambda_V \approx \lambda_A$, so far we have not produced any deeper reason for such a connection within the framework of model 1). We know only that $\lambda_V = \lambda_A = 1$ in the limit of complete $SU(3)$-symmetry.

7. COMMUTATORS AND PARTIALLY CONSERVED CURRENTS

A possible source for information about the relative magnitude of current matrix elements are the commutation relations of these currents as generators of $U_L(6) \times U_R(6)$ and its subgroups, or even of $U(12)$. However, in these commutation relations we must introduce a complete set of physical states in order to obtain expressions involving the products of matrix elements of individual currents.[35] The equations obtained in this way are directly useful only if we can restrict the intermediate states to those with one or two particles. In general, there is no reason why such approximations should make sense. In the decomposition of the commutators there are *no* energy denominators, as in the case of dispersion representations, where they may give rise to some suppression of higher mass states.

The situation may be more favorable if we consider commutators involving the divergence of axial vector currents and if we assume that these currents are partially conserved. We write

$$\partial_\alpha A_{i\alpha}(x) = C_i P_i(x) \tag{90}$$

where the $P_i(x)$ are pseudoscalar densities. We assume that these are essentially given by the field operators of the corresponding pseudoscalar particles,[16] that is, π-mesons and K-mesons for $i = 1 \ldots 7$. Of course, there are correction terms, but we propose that their matrix elements can be neglected as a first approximation. As is well known, this approximation leads directly to the G.T. relations.[36] However, in the way in which we use it here, the smallness of the correction term is more critical because of the absence of a damping energy denominator.

In terms of the basic triplet fields introduced in Section 3, we can write pseudoscalar and scalar densities in the form

$$\begin{aligned} P_i(x) &= i\bar{\varphi}(x)\gamma_5 \frac{\lambda_i}{2} \varphi(x) \\ S_i(x) &= \bar{\varphi}(x) \frac{\gamma_i}{2} \varphi(x) \end{aligned} \tag{91}$$

Using equation (28), we obtain their commutation relations with the generators of $U_L(6) \times U_R(6)$ and its subgroups which may then be abstracted form the model. Here it is sufficient to consider the generators

$$F_{i0} = \int d^3x [j_{i0}^{(+)}(x) + j_{i0}^{(-)}(x)]$$

and

$$F_{i0}^5 = \int d^3x [j_{i0}^{(+)}(x) - j_{i0}^{(-)}(x)] \qquad i = 0, 1 \ldots 8$$

of the subgroup $U_L(3) \times U_R(3)$. We find the commutation relations for equal times[17]

$$\begin{aligned} [F_{i0}, S_j(x)] &= if_{ijk} S_k(x), \qquad & [F_{i0}^5, S_j(x)] &= -id_{ijk} P_k(x) \\ [F_{i0}, P_j(x)] &= if_{ijk} P_k(x), \qquad & [F_{i0}^5, P_j(x)] &= id_{ijk} S_k(x) \end{aligned} \tag{92}$$

In contrast to the densities $j_{10}(x)$ and $j^5_{10}(x)$, which transform according to the representation $(1, 8)$ and $(8, 1)$ of $U_L(3) \times U_R(3)$, the scalars and pseudoscalars belong to $(3, 3^*)$ and $(3^*, 3)$.

It is convenient to write down a Hamiltonian for the triplet model in order to specify the transformation properties of the symmetry-breaking terms. As a special example, we consider

$$H = H_{\text{inv}} + \beta H_0 + \gamma H_8 \tag{93}$$

where H_0 and H_8 transform according to S_0 and S_8, respectively, and H_{inv} is invariant under $U_L(3) \times U_R(3)$. Instead of a $SU(3)$-breaking term belonging to $(3, 3^*)$ and $(3^*, 3)$, we could also use a corresponding member of the representation $(1, 8)$ and $(8, 1)$. The "mass term" H_0 commutes with the generators F_i of $SU(3)$, but it is not γ_5-invariant.

We can determine the relevant coefficients C_1 in equation (90) using the familiar relation

$$\dot{F}^5_{i0} = i[H, F^5_{i0}] \tag{94}$$

and recalling that

$$\dot{F}^5_{i0}(x_0) = \int d^3x \partial_\alpha A_{i\alpha}(x) \tag{95}$$

The result is

$$\begin{aligned} \partial_\alpha(A_{1\alpha} + iA_{2\alpha}) &= -\left(\beta\sqrt{\frac{2}{3}} + \gamma\sqrt{\frac{1}{3}}\right)(P_1 + iP_2) \\ \partial_\alpha(A_{4\alpha} + iA_{5\alpha}) &= -\left(\beta\sqrt{\frac{2}{3}} - \gamma\sqrt{\frac{1}{12}}\right)(P_4 + iP_5) \end{aligned} \tag{96}$$

Let us consider now those commutation relations which involve only axial-vector current.[8,31] At first, there are the relations which are related to $SU(2)$ subgroups of I-spin and V-spin. They are, with $F^5_{i0}(t) = F^5_{i0}$ and $t = x_0$,

$$\begin{aligned} [F^5_{10} - iF^5_{20}, \partial_\alpha(A_{1\alpha}(x) + iA_{2\alpha}(x))] &= \left(\beta\sqrt{\frac{2}{3}} + \gamma\sqrt{\frac{1}{3}}\right) \\ &\times \left\{\frac{1}{2} S_3(x) + \sqrt{\frac{2}{3}} S_0(x) + \sqrt{\frac{1}{3}} S_8(x)\right\} \end{aligned} \tag{97}$$

$$\begin{aligned} [F^5_{40} - iF^5_{50}, \partial_\alpha(A_{i\alpha}(x) + iA_{5\alpha}(x))] &= \left(\beta\sqrt{\frac{2}{3}} - \gamma\sqrt{\frac{1}{12}}\right) \\ &\times \left\{\frac{1}{2} S_3(x) + \sqrt{\frac{2}{3}} S_0(x) - \sqrt{\frac{1}{12}} S_8(x)\right\} \end{aligned} \tag{98}$$

But there are also the mixed commutators

$$[F^5_{10} - iF^5_{20}, \partial_\alpha(A_{4\alpha}(x) + iA_{5\alpha}(x))]$$
$$= -\left(\beta\sqrt{\frac{2}{3}} - \gamma\sqrt{\frac{1}{12}}\right)\frac{1}{2}(S_6(x) + iS_7(x)) \tag{99}$$

$$[F^5_{40} - iF^5_{50}, \partial_\alpha(A_{1\alpha}(x) + iA_{2\alpha}(x))]$$
$$= -\left(\beta\sqrt{\frac{2}{3}} + \gamma\sqrt{\frac{1}{3}}\right)\frac{1}{2}(S_6(x) - iS_7(x)) \tag{100}$$

they involve I-spin and V-spin components and they give rise to neutral, strangeness-changing densities.

In order to extract some information from equations (97) and (98), we can take the vacuum expectation values on both sides of these equations and introduce complete sets of intermediate states. In the approximation where the divergences $\partial_\alpha(A_{1\alpha} + A_{2\alpha})$ and $\partial_\alpha(A_{4\alpha} + iA_{5\alpha})$ act like π^-- and K^--field operators, respectively, we see that in the first terms of the commutators only $\pi^+(K^+)$ states and in the second terms only $\pi^-(K^-)$ states contribute. Using the definitions (63) for the quantities $B_\pi(m_\pi^2)$ and $B_K(m_K^2)$, we obtain the ratio

$$\frac{B_K^2}{B_\pi^2} \approx \frac{m_\pi^2}{m_K^2}\,\frac{\beta - \gamma/2\sqrt{2}}{\beta + \gamma/\sqrt{2}}\,\frac{1 - \rho/2\sqrt{2}}{1 + \rho/\sqrt{2}} \tag{101}$$

where we have used the definition

$$\rho = \frac{\langle 0|S_8(0)|0\rangle}{\langle 0|S_0(0)|0\rangle} \tag{102}$$

and the fact that $\langle 0|S_3(0)|0\rangle = 0$. Note that $\rho = 0$ in the limit of $SU(3)$-symmetry. Next we take the formulae (99) [and (100)] between the states $\langle\pi^-|$ and $|K^-\rangle$ [$\langle K^-|$ and $|\pi^-\rangle$]. Again, handling the divergence as a field operator, we can insert, in the first term of the commutator in (99), the intermediate states $|K^-K^+\rangle$ [$|\pi^+\pi^-\rangle$], whereas the second term requires the states $|K^-\pi^-\rangle$ [$|K^-\pi^-\rangle$]. We obtain rather involved relations between matrix elements like $\langle K^-\pi^-|A_{10} - iA_{20}|K^-\rangle$, etc., which are of no direct use for our purpose.

If we assume that $\gamma \ll \beta$, which also should imply $\rho \ll 1$, we obtain for equation (101) the desired reduction factor for strangeness changing transitions:

$$\frac{B_K}{B_\pi} \approx \frac{m_\pi}{m_K}$$

We turn now to the commutation relations involving the vector currents in the hope to get some information concerning the other problem of model 1)[31,37]: why is the damping factor λ_V of the same magnitude as λ_A? Unfortunately, there seems to be no simple answer to this question.

First, we consider again the commutation relations related for the I-spin and V-spin subgroups:

$$[F_{10} - iF_{20}, \partial_\alpha[A_{1\alpha}(x) + iA_{2\alpha}(x)]] = \left(\beta\sqrt{\frac{2}{3}} + \gamma\sqrt{\frac{1}{3}}\right)P_3(x) \tag{103}$$

$$[F_{40} - iF_{50}, \partial_\alpha[A_{4\alpha}(x) + iA_{5\alpha}(x)]] = \frac{1}{2}\left(\beta\sqrt{\frac{2}{3}} - \gamma\sqrt{\frac{1}{12}}\right)[P_3(x) + \sqrt{3}\,P_8(x)] \tag{104}$$

It is perhaps helpful to give some details of the evaluation of these equations. We take the matrix elements between the states $\langle\pi^0|$ and $|0\rangle$ and insert the intermediate states which are allowed by our hypothesis of a partially conserved axial-vector current. Then we obtain from equation (104), as an example,

$$\int\frac{d^3K}{2K_0}\delta(\bar{K} - \bar{\pi})\langle\pi^0|V_{40}(0) - iV_{50}(0)|K^+\rangle\langle K^+|\partial_\alpha(A_{4\alpha} + iA_{5\alpha})(0)|0\rangle$$
$$- \int\frac{d^3K}{2K_0}\int\frac{d^3\pi'}{2\pi'_0}\delta(\bar{K} + \bar{\pi}'_0)\langle\pi^0|\partial_\alpha(A_{4\alpha} + iA_{5\alpha})(0)|\pi'^0K^-\rangle$$
$$\times\langle\pi^{0\prime}K^-|V_{40}(0) - iV_{50}(0)|0\rangle = \frac{1}{2}\left(\beta\sqrt{\frac{2}{3}} - \gamma\sqrt{\frac{1}{12}}\right)\langle\pi^0|P_3(0)|0\rangle \tag{105}$$

In our approximation, we have

$$\langle\pi^0|\partial_\alpha(A_{4\alpha} + iA_{5\alpha})|\pi^{0\prime}K^-\rangle \approx 2\pi'_0\delta(\bar{\pi} - \bar{\pi}')\langle -K^+| \times\partial_\alpha(A_{4\alpha} + iA_{5\alpha})|0\rangle \tag{106}$$

and using equations (63) and (67), we find for the left-hand side of equation (105)

$$\frac{1}{2K_0}\langle\pi^0|V_{40} - iV_{50}|K^+\rangle\langle K^+|\partial_\alpha(A_{4\alpha} + iA_{5\alpha})|0\rangle\Big|_{\bar{\pi}=\bar{K}}$$
$$- \frac{1}{2K_0}\langle -K^+|\partial_\alpha(A_{4\alpha} + iA_{5\alpha})|0\rangle\langle\pi^0|V_{40} - iV_{50}|-K^+\rangle\Big|_{\bar{\pi}=-\bar{K}}$$
$$= \frac{1}{\sqrt{2}}F_{\pi K}\left(\frac{K_0 + \pi_0}{2K_0} + \frac{K_0 - \pi_0}{2K_0}\xi_{\pi K}\right)m_K^2B_K$$
$$- \frac{1}{\sqrt{2}}F'_{\pi K}\left(\frac{-K_0 + \pi_0}{2K_0} + \frac{-K_0 - \pi_0}{2K_0}\xi'_{\pi K}\right)m_K^2B_K$$
$$= \frac{1}{2}\left(\beta\sqrt{\frac{2}{3}} - \gamma\sqrt{\frac{1}{12}}\right)\langle\pi^0|P_3|0\rangle \tag{107}$$

Here the form factors $F'_{\pi K}$, $\xi_{\pi K}$ and $F'_{\pi K}$, $\xi'_{\pi K}$ are evaluated at the arguments

$$(\pi_0 \pm K_0)^2$$

respectively, which correspond to $(m_K \pm m_\pi)^2$ for $\pi = 0$. If we assume that the variation of these form factors can be neglected in a first approximation, we find

$$\frac{1}{\sqrt{2}} F_{\pi K}(1 + \xi_{\pi K}) m_K^2 B_K \approx \frac{1}{2}\Big(\beta\sqrt{\frac{2}{3}} - \gamma\sqrt{\frac{1}{12}}\Big)\langle \pi^0 | P_3 | 0 \rangle \tag{108}$$

In equation (103) we also take the matrix element between the neutral pion state and the vacuum, and with the intermediate states $|\pi^+\rangle$ and $|\pi^0\pi^-\rangle$, we obtain the relation

$$\sqrt{2}\, F_{\pi\pi} m_\pi^2 B_\pi \approx \Big(\beta\sqrt{\frac{2}{3}} + \gamma\sqrt{\frac{1}{3}}\Big)\langle \pi^0 | P_3 | 0 \rangle \tag{109}$$

where $F_{\pi\pi} = F_{\pi\pi}(0) = 1$ as a consequence of the conserved current. The ratio of equations (108) and (109) gives

$$\frac{(1 + \xi_{\pi K})\, F_{\pi K} B_K}{F_{\pi\pi} B_\pi} \approx \frac{m_\pi^2}{m_K^2} \frac{\beta - \gamma/2\sqrt{2}}{\beta + \gamma/\sqrt{2}} \tag{110}$$

and together with equation (101), this formula would give the desired reduction factor $\lambda_v \approx m_\pi/m_K$ for $\gamma \ll \beta$ and $|\xi| \ll 1$.

We recall that $\xi_{\pi K}$ vanishes in the limit of SU (3)-symmetry. However, before we can discuss the relevance of this result, we have to evaluate the consequences of the mixed commutation relations

$$\begin{aligned} &[F_{10} - iF_{20}, \partial_\alpha (A_{4\alpha}(x) + iA_{5\alpha}(x))] \\ &\quad = \frac{1}{2}(\beta\sqrt{2/3} - \gamma\sqrt{1/12})(P_6(x) + iP_7(x)) \end{aligned} \tag{111}$$

$$\begin{aligned} &[F_{40} - iF_{50}, \partial_\alpha (A_{1\alpha}(x) + iA_{2\alpha}(x))] \\ &\quad = \frac{1}{2}(\beta\sqrt{2/3} + \gamma\sqrt{1/3})(P_6(x) - iP_7(x)) \end{aligned} \tag{112}$$

In equation (111) we take the matrix element between $\langle K^0|$ and $|0\rangle$, and in equation (112) between $\langle \bar{K}^0|$ and $|0\rangle$. The intermediate state are $|K^-\rangle$ and $|\bar{K}^0 K^+\rangle$ in the first equation, and $|\pi^-\rangle$ and $|K^0\pi^+\rangle$ in the secone one. With these we obtain the relations

$$F_{KK} m_K^2 B_K = \frac{1}{2}(\beta\sqrt{2/3} - \gamma\sqrt{1/12})\langle K^0 | P_6 + iP_7 | 0 \rangle \tag{113}$$

where again $F_{KK} = F_{KK}(0)$, and

$$F_{\pi K}\left(\frac{\pi_0 + K_0}{2\pi_0} + \frac{K_0 - \pi_0}{2\pi_0}\xi_{\pi K}\right)m_\pi^2 B_\pi$$

$$- F'_{\pi K}\left(\frac{-\pi_0 + K_0}{2\pi_0} + \frac{+K_0 + \pi_0}{2\pi_0}\xi\pi_{\pi' K}\right)m_\pi^2 B_\pi$$

$$= \frac{1}{2}(\beta\sqrt{2/3} + \gamma\sqrt{1/3})\langle\bar{K}^0|P_6 - iP_7|0\rangle \tag{114}$$

The argument of the form factors are $(\pi_o - K_o)^2$ and $(\pi_o + K_o)^2$ for unprimed and primed functions, respectively. If we ignore their variation, the formula (114) reduces to

$$F_{\pi K}(1 - \xi_{\pi K})B_\pi m_\pi^2 \approx \frac{1}{2}(\beta\sqrt{2/3} + \gamma\sqrt{1/3})\langle\bar{K}^0|P_6 - iP_7|0\rangle \tag{115}$$

and in the ratio with equation (113) we have

$$\frac{(1 - \xi_{\pi K})F_{\pi K}B_\pi}{F_{KK}B_K} \approx \frac{m_K^2}{m_\pi^2} \cdot \frac{\beta + \gamma/\sqrt{2}}{\beta - \gamma/2\sqrt{2}} \tag{116}$$

We emphasize that the formulas (108) and (115) have been obtained from equations (107) and (114) by assuming *only* that the form factors $F_{\pi K}$ and $\xi_{\pi K}$ are slowly varying functions. No approximations have been made concerning the kinematic factors. It should be pointed out also, that our handling of the divergence of the axial-vector currents as field operators corresponds to an approximation which is quite different from a simple, one-particle approximation. Furthermore, we have not made a perturbation expansion in terms of the $SU(3)$-breaking part of the strong interactions.

The assumption that the form factor $F_{\pi K}(-q^2)$ is a slowly varying function of q^2 may not be very good for the interval $(m_K - m_\pi)^2 \leq -q^2 \leq (m_K + m_\pi)^2$. As an alternative, let us consider the possibility that $F_{\pi K}$ and $\xi_{\pi K}$ are approximately given by the K^*-meson dominance model. Then we have

$$F_{\pi K}(-q^2) = \frac{g_{\pi K}^* m^{*2}}{q^2 + m^{*2}}$$

and

$$\xi_{\pi K}(-q^2) = -\frac{m_K^* - m_\pi^2}{m^{*2}} \tag{117}$$

where $m^{*2} \approx 0.78\ (\text{BeV})^2$. Substituting these expressions into equation (107) and (114), and assuming $\rho \ll 1$ in equation (101), we find the relations

$$g^*_{\pi K} = 1 \qquad B_\pi = B_K \tag{118}$$

The equations (118) give an indication that the full current algebra implies that renormalization effects are small, in spite of the large πK-mass splitting.

Other approximations and different commutation relations will be discussed elsewhere.

Taking the ratios (110) and (116) together, and ignoring again the q^2-dependence of the form factors we obtain the relation

$$F^2_{\pi K}(1 - \xi^2_{\pi K}) \approx F_{\pi\pi} F_{KK} = 1 \tag{119}$$

because c.v.c. requires $F_{KK}(0) = F_{\pi\pi}(0) = 1$. There is no indication for a damping of the strangeness-changing decays, although the commutation relations (103) and (104) by themselves gave a result in this direction. In fact, if we take the equations (101), (110), and (116) together, we see that they have no solution for small values of γ, ρ, and $\xi_{\pi K}$. They are consistent only in the strict limit of $SU(3)$-symmetry, where $m_\pi = m_K$. Some caution is required in taking this limit, because we have used the approximation

$$F_{\pi K}[(K_o - \pi_o)^2] \approx F_{\pi K}[(K_o + \pi_o)^2] \qquad \text{with} \qquad K_o^2 = m_K^2 + \bar{\pi}^2$$

However, we can go back to equations (107) and (114). It appears that the notion of selected larger manifestations of the symmetry-breaking interaction is not compatible with the commutation relations (103) and (104) and the rather restrictive formulation of the partially conserved axial-vector current hypothesis which we have used. If we relax this hypothesis, higher mass states will come into play, and it becomes difficult to make predictions. In fact, in the limit where high mass states are dominant, we may well expect that renormalization effects are rather small.

Instead of neglecting the variation of the invariant form factors in equations (107) and (114), we could take the limit $\bar{\pi}^2 \to \infty$. Then we obtain equations (110) and (116) with $(1 \pm \xi_{\pi K})F_{\pi K}$ replaced by

$$F_{\pi K}(0) \pm \xi_{\pi K}(\infty) F_{\pi K}(\infty)$$

and hence equation (117) becomes

$$F^2_{\pi K}(0) \sim 1 + \xi^2_{\pi K}(\infty) F^2_{\pi K}(\infty)$$

This relation indicates $F_{\pi K}(0) \gtrsim 1$ or $F_{\pi K}(0) \approx 1$, because the form factor presumably vanishes for $q \to \infty$.

Although it is difficult in general, to envisage that the commutation relations (103) and (104) have a different domain of validity than the commutators (111) and (112), it is perhaps of interest that the first set is analogous to the relations $[I_+, I_-] = 2I_3$ and $[V_+, V_-] = I_3 + 3Y/2$, whereas, the second set is associated with the mixed commutators $[I_-, V_+] = U_+$ and $[I_+, V_-] = -U_-$ which give rise to neutral, strangeness-changing currents. If there should be some reason why the hypothesis of partially-conserved vector currents is a reasonable approximation only within the framework of the $SU(2)$-subgroups of I- and V-spin, then our equations (101) and (110) would give a nice dynamical explanation for a common reduction factor $\lambda_V \approx \lambda_A \approx m_\pi/m_K$. However, at present there seem to be no convincing arguments for such a situation, and we must conclude that the chiral $U(3) \times U(3)$ algebra indicates that the suppression of $\Delta S = 1$ transitions cannot be due to strong interaction effects.

ACKNOWLEDGMENTS

It is a pleasure to thank Professor L. Van Hove for his kind hospitality at CERN, where part of this work has been done. I am also very grateful for the hospitality of Professor A. Ramakrishnan and his group at the Summer School on Theoretical Physics in Bangalore, India. This work has been supported in part by the John Simon Guggenheim Foundation and by the United States Atomic Energy Commission.

REFERENCES

[1] R. Feynman and M. Gell-Mann, *Phys. Rev.* **109**: 193 (1958); E.C.G. Sudarshan and R.E. Marshak, *Proceedings of the Padua-Venice Conference*, September 1957, and *Phys. Rev.* **109**: 1860 (1958).

[2] G. Bernardini, *Proceedings of the 1965 Dubna Conference on High Energy Physics* (to be published).

[3] J. Schwinger, *Ann. of Physics* (New York) **2**: 407 (1957); A. Salam and J. C. Ward, *Nuovo Cimento* **11**: 568 (1959); *Phys. Letters* **13**: 168 (1964); S. Glashow, *Nuclear Phys.* **10**: 1013 (1959).

[4] M. Gell-Mann and Y. Ne'eman, *Ann. of Physics* (New York) **30**: 360 (1964).

[5] R. P. Feynman, M. Gell-Mann, and G. Zweig, *Phys. Rev. Letters* **13**: 678 (1964).

[6] B. Aubert, A. Behr, M. Block, J. P. Lowys, P. Mittner, and A. Ortin-Lecourtois, *Proc. of the Sienna International Conf. on Elementary Particles* (Italian

Physical Society, Bologna, 1963), p. 39; L. Kirsch, R. J. Plano, J. Steinberger, and P. Franzini, *Phys. Rev. Letters* **13**: 35 (1964); A. Burnstein, T. B. Day, A. J. Herz, B. Kehoe, B. Sechi-Zorn, N. Seeman, G. A. Snow, H. Courant, H. Filthuth, P. Franzini, R. G. Glasser, A. Minguzzi-Ranzi, A. Segar, and W. Willis, *Proc. of the International Conf. on Fundamental Aspects of Weak Interactions* (Brookhaven National Laboratory, Upton, New York, 1963), p. 427.

[7] J. H. Christenson, J.W. Cronin, U.L. Fitch, and R. Turlay, *Phys. Rev. Letters* **13**: 138 (1964).

[8] M. Gell-Mann, *Phys. Rev.* **125**: 1067 (1962).

[9] M. Gell-Mann, *Phys. Letters* **8**: 214 (1964).

[10] G. Zweig, CERN Report No. 8419/TH. 412 (1964).

[11] K. Bardacki, J. M. Cornwall, P. G. Freund, and B. W. Lee, *Phys. Rev. Letters* **13**: 698 (1964); R.E. Marshak and S. Okubo, *Phys. Rev. Letters* **13**: 818 (1964); R. Delbourgo, A. Salam, and J. Strathdee, Preprint ICTP 64/7 (to be published).

[12] F. Gürsey and L. Radicati, *Phys. Rev. Letters* **13**: 173 (1964); A. Pais, *Phys. Rev. Letters* **13**: 175 (1964); B. Sakita, *Phys. Rev.* **136**: B1756 (1964).

[13] M. Gell-Mann, *Phys. Rev. Letters* (to be published).

[14] M. Gell-Mann, Triplets and Triality, preprint CALT-68-8 (1964); M. Nauenberg, *Phys. Rev.* **135**: B1042 (1964); R. J. Oakes, *Phys. Rev.* **132**: 2349 (1963).

[15] C. Lovelace, Imperial College Preprint ICTP 64/66 (1964).

[16] M. Gell-Mann and M. Levy, *Nuovo Cimento* **16**: 705 (1960); M. Gell-Mann, *Proceedings of the 1960 Annual International Conf. on High Energy Physics at Rochester* (Interscience, New York, 1960), p. 508.

[17] M. Gell-Mann, *Physics* **1**: 63 (1964).

[18] N. Cabibbo, *Phys. Rev. Letters* **10**: 531 (1963).

[19] B. d'Espagnat and J. Prentki, *Nuovo Cimento* **24**: 497 (1962).

[20] J. J. de Swart, *Revs. of Mod. Phys.* **35**: 916 (1963).

[21] N. Cabbibo and R. Gatto, *Nuovo Cimento* **21**: 872 (1961).

[22] N. Brene, B. Hellesen, and M. Roos, *Phys. Letters* **11**: 344 (1964).

[23] R. H. Dalitz, *Properties of the Weak Interactions* (Varenna Lectures, 1964), Oxford preprint.

[24] S. Weinberg, *Phys. Rev.* **112**: 1375 (1958); N. Cabibbo, *Phys. Letters* **12**: 137 (1964).

[25] Dalitz, *op. cit.*; A. Pais and S. B. Treiman, *Proceedings of the 1964 Dubna Conference on High Energy Physics* (to be published).

[26] R. P. Feynman, *Proceedings of the 1960 Annual International Conf. on High Energy Physics at Rochester* (Interscience New York, 1960), p. 501.

[27] W. Willis *et al.*, *Phys. Rev. Letters* **13**: 291 (1964).

[28] A. W. Martin and K. C. Wali, *Phys. Rev.* **130**: 2455 (1963).

[29] M. Ademollo and R. Gatto, *Phys. Rev. Letters* **13**: 264 (1964); C. Bouchiat and P. Meyer, CERN preprint No. 9321 (1964).

[30] S. Fubini and G. Furlan, CERN preprint No. 9832/TH 487 (1964).

[31] R. Oehme, *Phys. Rev. Letters* **12**: 550, 604 (E) (1964).

[32] Riazuddin, *Phys. Rev.* **136**: 268 (1964).

[33] M. L. Goldberger and S. B. Theiman, *Phys. Rev.* **111**: 358 (1958).

[34] B. Barrett and T. N. Truong, *Phys. Rev. Letters* **13**: 734 (1964).

[35] W. Kummer, H. Pietschmann, and A. Balachandran, *Ann. of Physics* (New York) **29**: 161 (1964); K. Kawarabayashi and W. W. Wada, *Phys. Rev.* (to be published).

[36] J. Bernstein, S. Fubini, M. Gell-Mann, and W. Thirring, *Nuovo Cimento* **17**: 757 (1960); Y. Nambu, *Phys. Rev. Letters* **4**: 380 (1960).

[37] R. Oehme and G. Segrè, *Phys. Letters* **11**: 94 (1964).

[38] Chai S. Lai, University of Chicago (unpublished).

Partial Muon Capture in Light Nuclei

A. FUJII*

SOPHIA UNIVERSITY
Tokyo, Japan

In this lecture series, I will cover the following:

1. Atomic capture rate in hydrogen.
2. Molecular processes and molecular capture rate in liquid hydrogen.
3. Capture rate in He^3.
4. Partial capture rate in nuclei.
5. Experimental partial capture rate.

The elementary process of muon capture is the reaction

$$\mu^- + p \rightarrow n + \nu_\mu$$

which is quite analogous to the K-electron capture channel

$$e^- + p \rightarrow n + \nu_e$$

of beta decay. However, the released energy in muon capture is much higher than in beta decay so that the effect of the strong interaction is expected to be emphasized, while beta decay reaction merely reveals a threshold property.

The elementary process is not observable directly in the laboratory, at the present time, due to the limitation of the muon beam intensity. The molecular capture in liquid hydrogen and the partial capture in light nuclei, however, can provide the information of the elementary process indirectly.

---

*Department of Physics.

In this report the theoretical and experimental data are reviewed, with the aim of trying to find whether the picture of the universal $V-A$ law holds in the muon capture reaction or not.

1. ATOMIC CAPTURE RATE IN HYDOGEN

In the universal $V-A$ theory the S-matrix element of the elementary process

$$\mu^-(K\text{-orbit}) + p \rightarrow n + \nu$$

is written in a form

$$\langle n\nu|S|p\mu\rangle = -i\frac{G}{\sqrt{2}}\frac{1}{\sqrt{(2\pi)^9}}\frac{1}{\sqrt{\pi a^3}}(2\pi)^4\,\delta\,(p+\mu-n-\nu)$$
$$\times\;\bar{u}\,(n)\,(V_\lambda + P_\lambda)\,u\,(p)\,\bar{u}\,(\nu)\,\gamma_\lambda\,(1+\gamma_5)\,u\,(\mu) \qquad (1)$$

where G is the Fermi coupling constant, a is the Bohr radius of the mu-mesic atom ($p\mu$), p, n, μ, and ν stand for the four-momentum of the proton, neutron, muon, and neutrino, respectively. V_λ and P_λ are the vector and axial vector operators defined by

$$V_\lambda = f_1\,(k^2)\,\gamma_\lambda - \frac{f_2\,(k^2)}{m}\gamma_{\lambda\rho}\,k_\rho \qquad f_1\,(0) = 1 \qquad (2)$$

$$P_\lambda = g_1\,(k^2)\,\gamma_\lambda\,\gamma_5 - \frac{ig_2\,(k^2)}{m}\,k_\lambda\,\gamma_5 \qquad (3)$$

where k is the four-momentum transfer

$$k = n - p = \mu - \nu \qquad (4)$$

m is the nucleon mass and

$$\gamma_{\lambda\rho} \equiv \frac{1}{2i}(\gamma_\lambda\,\gamma_\rho - \gamma_\rho\,\gamma_\lambda)$$

This form of V_λ and P_λ follows from the assumption of a) Lorentz-invariance and b) definite G-conjugation parity

$$G\,V_\lambda\,G^{-1} = V_\lambda \qquad G\,P_\lambda\,G^{-1} = -P_\lambda \qquad (5)$$

The form factors f and g can be chosen all real if c) time-reversal invariance holds.

There is no experimental evidence for or against the definite G-parity in the weak current, but we may use it as a working hypothesis.

The relation between the effective coupling constant[1] and the form factors are:

$$C_V = \frac{G}{\sqrt{2}} f_1(k^2) \qquad \text{vector coupling constant}$$

$$-C_M = \frac{G}{\sqrt{2}} f_2(k^2) \qquad \text{magnetic coupling constant}$$

$$-C_A = \frac{G}{\sqrt{2}} g_1(k^2) \qquad \text{axial vector coupling constant}$$

$$C_P = \frac{G}{\sqrt{2}} \frac{m_\mu}{m} g_2(k^2) \qquad \text{pseudo-scalar coupling constant}$$

where the reaction takes place at $k^2 = m_\mu^2[1 - (m_\mu/m)]$.

The atomic capture rate is extremely sensitive to the relative orientation of the proton and muon spin (hyperfine effect). To illustrate this, we take the nonrelativistic approximation of the matrix element (1)

$$\langle n\nu|S|p\mu\rangle_{nr} = -i\frac{G}{\sqrt{2}}\frac{1}{\sqrt{(2\pi)^9}}\frac{1}{\sqrt{\pi a^3}}(2\pi)^4\,\delta(p+\mu-n-\nu)$$
$$\times\, \chi_n^+\,[f_1 - g_1(\boldsymbol{\sigma}_N\cdot\boldsymbol{\sigma})]\,\chi_p\,\frac{1}{\sqrt{2}}\chi_\nu^+\,(1-\boldsymbol{\sigma}\cdot\hat{\boldsymbol{\nu}})\,\chi_\mu \qquad (6)$$

where χ_α is the Pauli spinor for the particle α, $\boldsymbol{\sigma}_N$ and $\boldsymbol{\sigma}$ are the Pauli spin operators for nucleons and leptons, respectively, and $\hat{\boldsymbol{\nu}}$ is the unit vector along the three-momentum of the neutrino. The matrix element squared is proportional to

$$\begin{aligned} M &= |\,\chi_n^+\,[f_1 - g_1(\boldsymbol{\sigma}_N\cdot\boldsymbol{\sigma})]\,\chi_p|^2 \\ &= f_1^2 + g_1^2\langle(\boldsymbol{\sigma}_N\cdot\boldsymbol{\sigma})^2\rangle - 2f_1g_1\langle\boldsymbol{\sigma}_N\cdot\boldsymbol{\sigma}\rangle \\ &= f_1^2 + 3g_1^2 - 2(g_1^2 + f_1g_1)\langle\boldsymbol{\sigma}_N\cdot\boldsymbol{\sigma}\rangle \end{aligned} \qquad (7)$$

The expectation value of the relative spin orientation takes the value

$$\langle\boldsymbol{\sigma}_N\cdot\boldsymbol{\sigma}\rangle = \begin{cases} -3 & \text{for spin singlet} \\ +1 & \text{for spin triplet} \end{cases}$$

so that

$$M = \begin{cases} (f_1 + 3g_1)^2 & \text{for spin singlet} \\ (f_1 - g_1)^2 & \text{for spin triplet} \end{cases} \qquad (8)$$

The $V-A$ theory is characterized by $f_1 \sim g_1$, namely the triplet capture rate vanishes. This characteristic feature also continues to hold in the relativistic case, the triplet capture rate is some 50 times smaller than the singlet capture rate.

The expression of the relativistic capture rate was first given by Adams.[2] The form factors as functions of k^2 can be obtained by two dynamical principles: a) hypothesis of conserved vector current (c.v.c.) and b) hypothesis of asymptotically conserved axial vector current (a.c.a.c.).

The c.v.c. hypothesis predicts that the functional dependence of f_1 and f_2 are the same as the corresponding functions of the electromagnetic current, which are empirically known from the electron scattering experiment as

$$f_1(z) = -0.20 + \frac{1.20}{1 + 2.268z} \tag{9}$$

$$f_2(z) = \frac{\mu_p - \mu_n}{2} f_1(z) = 1.853\, f_1(z) \tag{10}$$

$$z \equiv \frac{k^2}{m^2}$$

The a. c. a. c. hypothesis assumes that the operator $\partial_\lambda P_\lambda$ is a gentle operator, which emphasizes the low frequency component and is practically a projection operator to the pion field. It gives a relation between g_1 and g_2 (extended Goldberger–Treiman relation):

$$2mg_1(k^2) - \frac{k^2}{m} g_2(k^2) = \frac{a_\pi g_\pi}{k^2 + m_\pi^2} + m\varphi(k^2) \tag{11}$$

where a_π is the pion decay amplitude, g_π is the renormalized symmetrical pion-nucleon coupling constant, and the function $\varphi(k^2)$ is a slowly-varying function in the vicinity of $|k|^2 \lesssim m_\pi^2$ with the property $\varphi(-m_\pi^2) = 0$. Assume that $g_1(k^2)$ can be expanded in power series around $k^2 = 0$,

$$g_1(z) = g_1(0)\left[1 - \frac{1}{6}\rho_A z + \dots\right] \tag{12}$$

where the parameter ρ_A defined in this equation may be called the "A-radius" of the nucleon. Then subtracting the equation for $k^2 = 0$ in equation (11) from equation (11) itself, we obtain

$$g_2(z) = \frac{a_\pi g_\pi}{mm_\pi^2} \frac{1}{z + (m_\pi/m)^2} - \frac{1}{3}\rho_A \cdot g_1(0) \tag{13}$$

where the condition $\varphi(k^2) \sim \varphi(0)$, $|k|^2 \lesssim m_\pi^2$ is used. The form factors are thus expressed by two parameters ρ_A and $g_1(0)$, but the universality tells that $g_1(0)$ is the ratio of the axial vector and vector coupling constant in nucleon beta decay. Under some arbitrary but plausible choice of ρ_A, we can compute the atomic capture rate from Adams' formula at $z = 0.0114$, namely,

$$k^2 = m_\mu^2 \left(1 - \frac{m_\mu}{m}\right)$$

2. MOLECULAR PROCESSES AND MOLECULAR CAPTURE RATE IN LIQUID HYDROGEN

Because of present day intensity of the muon beam, we must use a dense proton target as liquid hydrogen. The processes, before the muon disappears by decay or capture reaction by the proton, are summarized in Fig. 1. From the observed reaction rate, we compute the branching ratios in each state and summarize then in Fig. 2. A very small contamination of the deuteron plays a very important role by "snatching" the muon from the proton, baiting a smaller binding energy due to the larger reduced mass.

Reaction I $(p\mu)_{\text{triplet}} \rightarrow (p\mu)_{\text{singlet}} + 0.25\text{eV}$

II $(p\mu) + (pp) \rightarrow (p\mu p) + p + 90\text{eV}$

III $(p\mu) + (pd) \rightarrow (d\mu) + 2p + 134\text{eV}$

IV $(d\mu) + (pp) \rightarrow (d\mu p) + p + 93\text{eV}$

V $(d\mu p)_{\text{ortho}} \rightarrow (d\mu p)_{\text{para}} + 124\text{eV}$

Experimental Transition Rates

1. Experiment by E. Bleser *et al.*[3]

$$\lambda_{\text{II}} = (1.4 \pm 0.5) \times 10^6\ \text{sec}^{-1}$$

$$\lambda^*_{\text{III}} = (0.91 \pm 0.2) \times 10^6\ \text{sec}^{-1}$$

$$\lambda_{\text{IV}} = (5.5 \pm 1.1) \times 10^6\ \text{sec}^{-1}$$

λ^* being calculated by assuming the deuterium concentration 50 ppm.

2. Experiment by G. Conforto *et al.*[4]

$\lambda_{\text{II}} = (3.26 \pm 0.78) \times 10^6\ \text{sec}^{-1}$ and $(2.81 \pm 0.16) \times 10^6\ \text{sec}^{-1}$

$\lambda_{\text{IV}} = (6.55 \pm 0.46) \times 10^6\ \text{sec}^{-1}$, $(7.18 \pm 0.67) \times 10^6\ \text{sec}^{-1}$, and $(7.75 \pm 0.77) \times 10^6\ \text{sec}^{-1}$

Two or more experimental values of λ are deduced from the experiment done with different Ne-deuterium concentration.

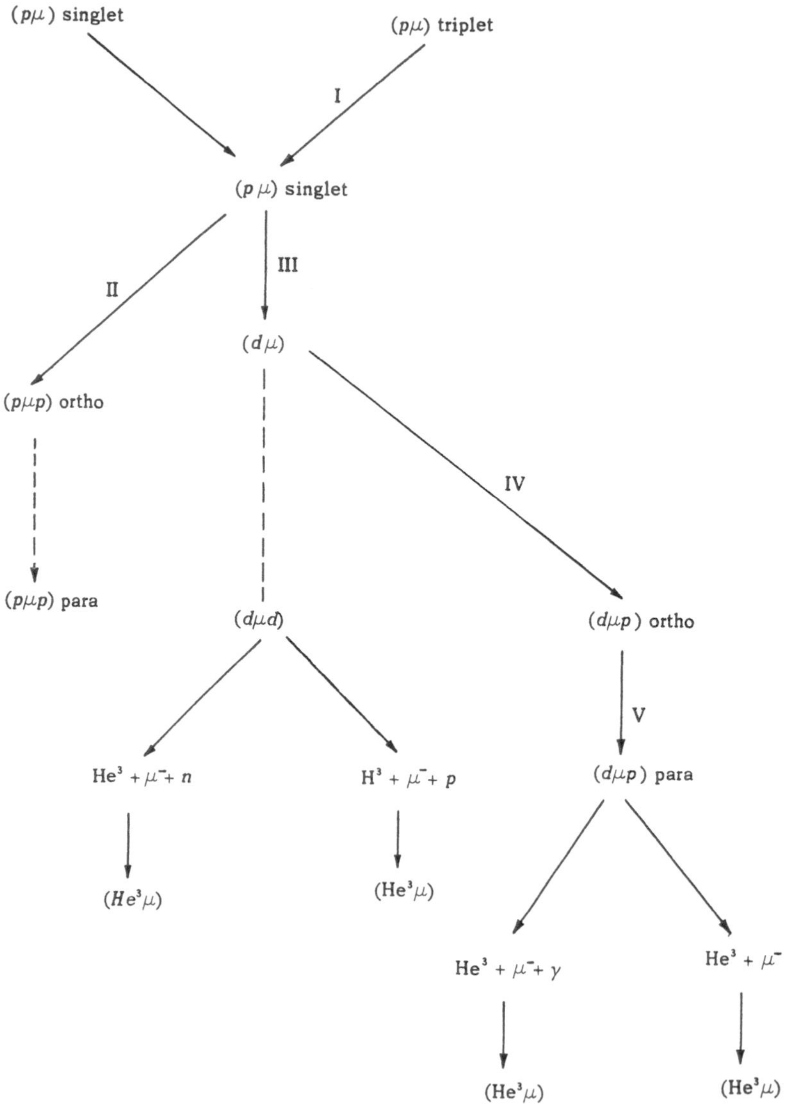

Fig. 1. Molecular process in liquid hydrogen.

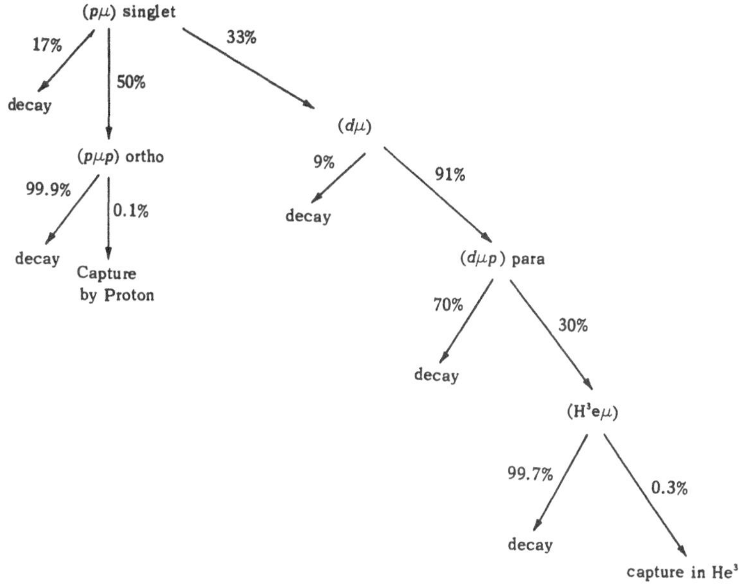

Fig. 2. Branching ratios of muon reaction in liquid hydrogen.

Theoretical Transition Rates

$\lambda_I = 10^9\ sec^{-1}$	Reference 5
$\lambda_{II} = 6.5 \times 10^6\ sec^{-1}$	6
$= 2.5 \times 10^6\ sec^{-1}$	7, 8, 9
$\lambda_{III} = 1.4 \times 10^{10}\ sec^{-1}$	7, 8
$1.3 \times 10^{10}\ sec^{-1}$	7, 6
$1.8 \times 10^{10}\ sec^{-1}$	7, 10
$\lambda_{IV} = 2.5 \times 10^6\ sec^{-1}$	6
$1.2 \times 10^6\ sec^{-1}$	7, 8
$\lambda_V = 2.5 \times 10^{10}\ sec^{-1}$	6

1. In compiling this diagram we used the experimental value of λ_{II} by Bleser *et al.* [Ref. 3, Fig. 1], muon decay rate $0.455 \times 10^6\ sec^{-1}$, capture rate in molecular ion $(p\mu p) \sim 500\ sec^{-1}$, capture rate in $He^3 \sim 1500\ sec^{-1}$.

2. The overall branching ratio is, from the diagram,

capture in proton	0.05%
capture in He^3	0.03%
decay	99.92%

We notice that 40% of the whole capture process is due to He^3 induced by the tiny (5×10^{-5}) concentration of the deuterium impurity.

3. The population of the final seat of the muon is, from the diagram,

$(p\mu)$	17%
$(p\mu p)$	50%
$(d\mu)$	3%
$(d\mu p)$	21%
$(He^3\mu)$	9%

As is shown in Fig. 2 the absorption of muon in hydrogen takes place from the ortho molecular ion $(p\mu p)$, which is characterized such that two proton spins are parallel and the muon spin is anti-parallel to them. The muon has no orbital angular momentum, and the total angular momentum of the molecular ion is $J = 1/2$. The molecular capture rate w is related to the hyperfine atomic capture rate w_0 (spin singlet) and w_1 (spin triplet) as

$$w = 2\gamma \left(\frac{3}{4} w_0 + \frac{1}{4} w_1\right)$$

The weights are just opposite to the statistical factor, but it comes from $J = 1/2$ as follows:

The capture rate is written in a form, explicit in the relative orientation of the proton and muon spin

$$w = a + b \langle \boldsymbol{\sigma}_N \cdot \boldsymbol{\sigma} \rangle$$

where a and b are constants, $\boldsymbol{\sigma}_N$ and $\boldsymbol{\sigma}$ are the proton and muon spin vectors, respectively. For the atomic capture rate we have

$$w_0 = a - 3b, \qquad w_1 = a + b$$

For the molecular ion, if $\mathbf{J}$ and $\mathbf{J}_{PP}$ represents the angular momentum of the whole molecule and the two-proton system,

$$\mathbf{J} = \mathbf{J}_{pp} + \frac{1}{2}\boldsymbol{\sigma}$$

$$\therefore \mathbf{J}^2 = \mathbf{J}_{pp}^2 + \frac{1}{4}\boldsymbol{\sigma}^2 + (\mathbf{J}_{pp} \cdot \boldsymbol{\sigma})$$

$$= \mathbf{J}_{pp}^2 + \frac{1}{4}\boldsymbol{\sigma}^2 + (\boldsymbol{\sigma}_N \cdot \boldsymbol{\sigma})$$

since the two protons are equivalently seen by the muon. We obtain

$$\langle \boldsymbol{\sigma}_N \cdot \boldsymbol{\sigma} \rangle = J(J+1) - J_{pp}(J_{pp}+1) - \frac{3}{4} = -2$$

and

$$w = a - 2b = \frac{3}{4} w_0 + \frac{1}{4} w_1$$

The other parameter γ is the correction of the muon overlapping with the proton. It is defined by the ratio of the muon density at the position of either proton in the ortho-molecular ion and the muon density at the position of the proton in the mu-mesic atom. Weinberg[11] estimated $2\gamma = 1.17$, *subject to* the correction of the order m_μ/m.

If we collect the working formulae in Section 1 and Section 2, which are to be inserted in Adams' formula,

$$\begin{cases} f_1(z) = -0.20 + \dfrac{1.20}{1 + 2.268z} = 0.9698 \qquad \text{at } z = 0.0114 \\ f_2(z) = 1.853 f_1(z) = 1.797 \\ g_1(z) = g_1(0)\left(1 - \dfrac{1}{6} p_A z\right) \\ g_2(z) = \dfrac{a_\pi g_\pi}{m m_\pi^2} \dfrac{1}{z + 0.0221} - \dfrac{1}{3} p_A g_1(0) = 64.18 - \dfrac{1}{3} p_A g_1(0) \end{cases}$$

$$\begin{cases} w_0, w_1 \text{ by Adams' formula} \\ w = 1.17\left(\dfrac{3}{4} w_0 + \dfrac{1}{4} w_1\right) \end{cases}$$

the calculated atomic and molecular capture rates[12] summarized in Table 1 are obtained for certain choices of the parameters p_A and $g_1(0)$. The ρ_V corresponding to $f_1(z)$ is $\rho_V = 16.33$.

Table 1
Atomic and Molecular Capture Rate in Hydrogen, Theoretical

$g_1(0)$	p_A	$-C_P/C_A$	w_1, sec^{-1}	w_0, sec^{-1}	w, sec^{-1}
1.20	0	7.5	12.9	661	584
1.20	ρ_V	7.1	11.3	641	566
1.20	$2\rho_V$	6.7	9.9	620	547

The capture experiment in liquid hydrogen was carried out at the University of Chicago, Cern, and Columbia University. Quoting Rubbia's compilation[13]

Chicago	$428 \pm 85\ \text{sec}^{-1}$
Cern	$450 \pm 50\ \text{sec}^{-1}$
Columbia	$\begin{cases} 515 \pm 85\ \text{sec}^{-1} \\ 464 \pm 42\ \text{sec}^{-1} \end{cases}$

the world average is

$$w = 445 \pm 42\ \text{sec}^{-1}$$

The calculated rate is some 20 percent higher than the experiment, although they are not inconsistent with the original picture taking into account the uncertainities of molecular parameters. Recent calculation[14] has shown that $J = 3/2$ component in the ortho-molecular ion is very small, so that the molecular uncertainty is just boiled down to γ. If we assume that γ is not too far from the true value and want to explain the data *only* by straightening the effective $-C_P/C_A$ (not by ρ_A), it comes up as high as 16.[13]

Unless we obtain a much more reliable estimate of γ, it is difficult to say definitely whether our picture of the universal V–A theory *with* definite G-parity is really consistent with the experiment or not.

3. CAPTURE RATE IN HELIUM 3

A particular channel of the muon capture in He^3

$$\mu^- + He^3 \rightarrow H^3 + \nu \tag{14}$$

is very much alike the muon capture in atomic hydrogen. In fact, in many respects (He^3, H^3) doublet has features of $(\bar{n}, \bar{p})$ doublet, as may be seen by the picture that He^3/H^3 can be regarded as a neutron/proton hole in He^4.

The matrix element of the capture reaction (14) is again given by (1), (2), (3), but the only difference is that f's and g's are now the form factors of (He^3, H^3) doublet, which we may call the "trion". The empirical electromagnetic forms factors[17] provide the complete determination of the weak vector form factors upon the c. v. c. hypothesis. We shall again expand $g_1(z)$ as

$$g_1(z) = g_1(0)\left[1 - \frac{1}{6}\rho_A z\right]$$

where $g_1(0) = 1.186 \pm 0.026$ is obtained from the ft -value of H^3.

The nonrelativistic approximation of the operator $\gamma_\lambda\gamma_5$ is the same as $\gamma_{\lambda\rho}k_\rho$, we may use the experimental electromagnetic radius of the Pauli form factor $f_2(z)$ as a reasonable guess for ρ_A. The a. c. a. c. hypothesis may not be valid in this case, because the anomalous threshold comes down to almost $z = 0$ and no pion pole ($z = -0.0221$) dominance is expected at $z = 0.014$. We made nonrelativistic and impulse approximation, evaluated it with the main component of the ground state ${}^{22}S_{1/2}$ with the partition (1/2, 1/2, 1/2) and found that $g_2(z)$ for the trion was 3 times $g_2(z)$ for the nucleon.

The recent experiment[18] measures the rate of (14) in 3% precision as $w = 1520 \pm 50\ \text{sec}^{-1}$. Table 2 is the theoretical prediction made in the above framework, where the value of $g_2(z)$ is chosen as a parameter, i. e.,

$$\begin{cases} f_1(z) = 0.7981 \\ f_2(z) = \frac{1}{2}\left[\mu\,(\text{He}^3) - \mu\,(\text{H}^3)\right] \times 0.8881 = -7.232 \\ g_1(z) = g_1(0)\left(1 - \frac{1}{6}\rho_A z\right) = 1.186 \times 0.8881 = -1.053. \end{cases}$$

and μ is the trion anomalous magnetic moment in units of *trion* mageton.

Table 2

Partial Muon Capture Rate $\mu + \text{He}^3 \rightarrow \text{H}^3 + \nu$, Theoretical

$\frac{m_\mu}{m}\,g_2(z)$	w, sec^{-1}
−40	1467
−30	1498
−20	1558
−10	1648
0	1766
+10	1914

The experimental value fits in the range $-18 \geqq (m_\mu/m) \times g_2(z) \geqq -36$, but $(m_\mu/m)\,g_2(z) = 3 \times 7.5$ gives an excellent fit as $w = 1528\ \text{sec}^{-1}$. The degree of uncertainties in the theoretical prediction of the muon capture rate in He^3 seems to be less than that in liquid hydrogen. The picture summarized in equations (1), (2) and (3) for the elementary process seems to have a strong support in He^3 experiment.

4. PARTIAL CAPTURE RATE IN NUCLEI

The negative muon trapped in the Coulomb field of the nucleus cascades down to the lowest K-orbit in a sufficiently short time and forms a mu-mesic atom. Two channels are open there, decay or the nuclear capture

$$\mu^- + (A, Z) \rightarrow (A, Z - 1) + \nu$$

Most of the energy released by the muon mass is carried away by the neutrino, and the excitation of the nucleus from (A, Z) to $(A, Z - 1)$ is practically less than 30 MeV showing a predominance around 15 MeV. The nuclear transitions are depicted schematically in Fig. 3.

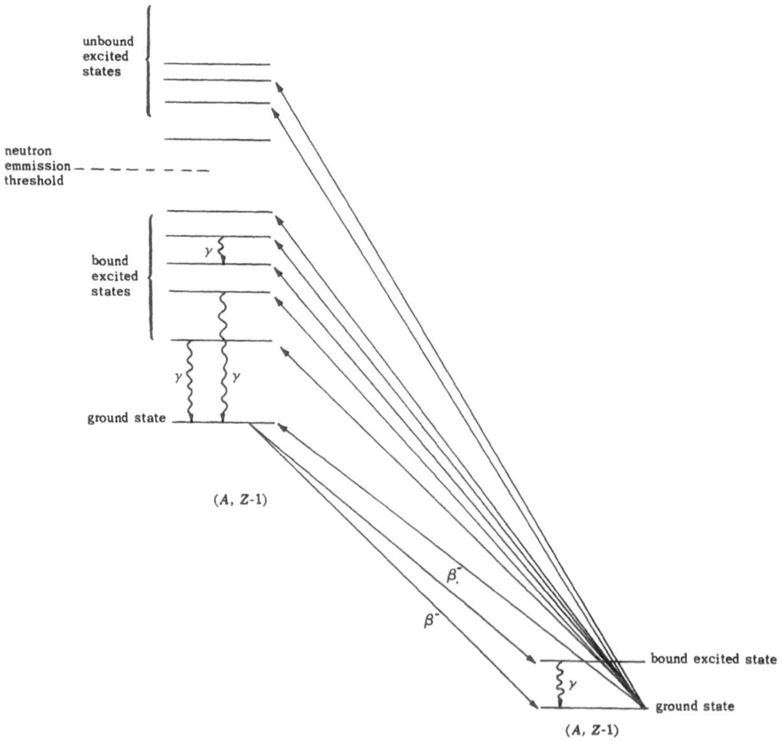

Fig. 3. Schematic description of nuclear excitation.

Many nuclear transitions are possible. If the final nuclear state is unbound, it decays into a different nuclide by emitting the neutron or other nuclear particles, while if the final state is bound it cascades down the ground state by emitting γ-ray. From the ground state β-decay usually takes place and the daughter nucleus $(A, Z-1)$ changes back to the parent (A, Z). The nonrelativistic impulse approximation is usually adopted. Namely, the nucleus is regarded as an assembly of the independent physical nucleons, and the probability that the muon is absorbed by the cluster of the nucleons is assumed negligible. The nonrelativistic approximation is necessary for computing the nuclear matrix element explicitly. The higher approximation in the nucleon velocity is possible, and obtained if necessary by the contact transformation of Foldy and Wouthuysen.[20]

The matrix element of the interaction Hamiltonian for the elementary process is

$$\langle n\nu|S|p\mu\rangle = \frac{G}{\sqrt{2}} N_\lambda L_\lambda \tag{15}$$

$$N_\lambda = \bar{u}(n)\left[f_1\gamma_\lambda - \frac{f_2}{m}\sigma_{\lambda\rho}\, k_\rho + g_1\gamma_\lambda\gamma_5 - \frac{ig_2}{m}\, k_\lambda\gamma_5\right] u(p) \tag{16}$$

$$L_\lambda = \bar{u}(\nu)\,\gamma_\lambda(1+\gamma_5)\,u(\mu) \tag{17}$$

the normalization constant being absorbed into Dirac spinors. The transcription of the operators (16) up to the first order of the nucleon momentum reads

$$\gamma_\lambda \to \begin{cases} -\dfrac{i}{2m}\boldsymbol{K} + \dfrac{1}{2m}(\boldsymbol{\sigma}\times\boldsymbol{k}) & \text{space part} \\ 1 & \text{time part} \end{cases}$$

$$\sigma_{\lambda\rho}k_\rho \to \begin{cases} -(\boldsymbol{\sigma}\times\boldsymbol{k}) \\ 0 \end{cases}$$

$$\gamma_\lambda\gamma_5 \to \begin{cases} i\boldsymbol{\sigma} \\ -\dfrac{1}{2m}(\boldsymbol{\sigma}\cdot\boldsymbol{K}) \end{cases}$$

$$k_\lambda\gamma_5 \to \frac{1}{2m}(\boldsymbol{\sigma}\cdot\boldsymbol{k})$$

where $\boldsymbol{K} = \boldsymbol{p} + \boldsymbol{n}$. The muon is assumed to have zero three-momentum. The lepton current is transcribed into Pauli spinors (no approximation for neutrino)

$$L_\lambda \rightarrow \begin{cases} \dfrac{1}{\sqrt{2}} \chi_\nu^+ (1 - \boldsymbol{\sigma} \cdot \hat{\boldsymbol{\nu}}) \cdot i\boldsymbol{\sigma}\, \chi_\mu & \text{space part} \\ \dfrac{1}{\sqrt{2}} \chi_\nu^+ (1 - \boldsymbol{\sigma} \cdot \hat{\boldsymbol{\nu}})\, \chi_\mu & \text{time part} \end{cases}$$

where $\hat{\boldsymbol{\nu}}$ is the unit vector along the three-momentum of the neutrino: $\boldsymbol{p}_\nu = p_\nu \hat{\boldsymbol{\nu}}$.

Hence,

$$\langle n\nu | S | p\mu \rangle = \frac{G}{2} \chi_\nu^+ \chi_n^+ H_{\text{eff}}\, \chi_p \chi_\mu \tag{18}$$

$$\begin{aligned} H_{\text{eff}} = (1 - \boldsymbol{\sigma} \cdot \hat{\boldsymbol{\nu}}) \Big\{ f_1 \left(1 + \frac{p_\nu}{2m}\right) 1_N \cdot 1 + \left[-g_1 - \frac{p_\nu}{2m}(f_1 + 2f_2) \right] \\ \times (\boldsymbol{\sigma}_N \cdot \boldsymbol{\sigma}) + \frac{p_\nu}{2m} \left[-(f_1 + 2f_2) + g_1 - \frac{m_\mu}{m} g_2 \right] (\boldsymbol{\sigma}_N \cdot \hat{\boldsymbol{\nu}}) + \frac{1}{m} f_1 (\boldsymbol{\sigma} \cdot \boldsymbol{p}) \\ - \frac{1}{m} g_1 (\boldsymbol{\sigma}_N \cdot \boldsymbol{p}) \Big\} \end{aligned} \tag{19}$$

where $\boldsymbol{\sigma}_N$, 1_N denotes the nucleon spin and unit operator and $\boldsymbol{p}$ is the nucleon momentum. Note that

$$(1 - \boldsymbol{\sigma} \cdot \hat{\boldsymbol{\nu}}) \cdot (\boldsymbol{\sigma} \cdot \hat{\boldsymbol{\nu}}) = -(1 - \boldsymbol{\sigma} \cdot \hat{\boldsymbol{\nu}})$$

so that $(\boldsymbol{\sigma} \cdot \hat{\boldsymbol{\nu}})$ can be replaced by -1 in equation (19).

Upon the assumption of impulse approximation the effective Hamiltonian for the nucleus, which gives the above matrix element for a single nucleon, is

$$\begin{aligned} H_{\text{eff}} = \frac{G}{2} \tau^+ (1 - \boldsymbol{\sigma} \cdot \hat{\boldsymbol{\nu}}) \sum_{i=1}^{A} \tau_i^{(-)} \delta(\mathbf{r} - \mathbf{r}_i) \\ \times \Big[G_V \cdot 1 \cdot 1_i + G_A (\boldsymbol{\sigma} \cdot \boldsymbol{\sigma}_i) - G_P (\boldsymbol{\sigma} \cdot \hat{\boldsymbol{\nu}}) (\boldsymbol{\sigma}_i \cdot \hat{\boldsymbol{\nu}}) \\ - g_V (\boldsymbol{\sigma} \cdot \hat{\boldsymbol{\nu}}) \left(\boldsymbol{\sigma}_i \cdot \frac{\boldsymbol{p}_i}{m} \right) - g_A (\boldsymbol{\sigma} \cdot \hat{\boldsymbol{\nu}}) \left(\boldsymbol{\sigma} \cdot \frac{\boldsymbol{p}_i}{m} \right) \Big] \end{aligned} \tag{20}$$

$$\begin{aligned} G_V &= f_1 \left(1 + \frac{p_\nu}{2m}\right) \\ G_A &= -g_1 - (f_1 + 2f_2) \frac{p_\nu}{2m} \\ G_P &= \left[-(f_1 + 2f_2) + g_1 - \frac{m_\mu}{m} g_2 \right] \frac{p_\nu}{2m} \\ g_V &= f_1 \\ g_A &= -g_1 \end{aligned} \tag{21}$$

where τ^+ is the lepton operator which transforms muon into neutrino and τ_i^- is the nucleon operator which transforms i-th proton into a neutron.

The partial capture rate between the initial nuclear state a and final b becomes

$$w_{ba} = \frac{1}{2\pi^3 a_z^3}\left(1 - \frac{p_\nu}{Am + m_\mu}\right) p_\nu^2 \frac{1}{2J_a + 1} |<b|H_{\text{eff}}|a>|^2$$

$$a_z = \frac{Z_{\text{eff}}}{137} \frac{m_\mu}{1 + \dfrac{m_\mu}{Am}} \tag{22}$$

$$p_\nu = (m_\mu - E_{ba})\left(1 - \frac{1}{2}\frac{m_\mu}{Am + m_\mu}\right)$$

where E_{ba} is the nuclear level distance between the states a and b. A tiny binding energy of the muon being neglected, J_a is the nuclear spin of the state a. The nuclear matrix element is characterized by the nuclear spin J and its component M:

$$|\langle b|H_{\text{eff}}|a\rangle|^2 \equiv |\langle J_b, M_b|H_{\text{eff}}|J_a, M_a\rangle|^2$$

$$= \int \frac{d\hat{\nu}}{4\pi} \left\{ G_V^2 \left| \int 1 \right|^2 + G_A^2 \left| \int \boldsymbol{\sigma} \right|^2 + (G_P^2 - 2G_P G_A) \left| \hat{\nu} \cdot \int \boldsymbol{\sigma} \right|^2 \right.$$

$$\left. + \text{ terms involving} \int p_i, \int (\boldsymbol{\sigma}_i \cdot p_i) \right] \Bigg\} \tag{23}$$

where

$$\int O \equiv \langle J_b, M_b| \sum_{i=1}^{A} \tau_i^{(-)} e^{-i\mathbf{p}_\nu \cdot \mathbf{r}_i} O \cdot \varphi_\mu(r_i)|J_a, M_a\rangle$$

$$O = 1, \boldsymbol{\sigma} \text{ etc.} \tag{24}$$

and $\varphi_\mu(r_i)$ is the radial part of the $1s$ muon orbital normalized as $\varphi_\mu(0) = 1$. If the nuclear wave function for the state a and b are known sufficiently, we can compute the nuclear matrix element and find the partial capture rate.

For a nucleus of finite size with a certain continuous distribution of the proton $\rho(x)$ the capture rate is proportional to

$$\langle \rho \rangle = \frac{\int_0^\infty \rho(x) |\psi_\mu(x)|^2 \, d^3x}{\int_0^\infty |\psi_\mu(x)|^2 \, d^3x}$$

We define the effective charge by

$$z_{\text{eff}}{}^4 = \langle \rho \rangle \, \pi a_0^3$$

where a_0 is the Bohr radius of the $(p\mu)$ atom. In other words, the square of the overlap of the muon, with the whole protons in terms of a wave function of the point charge Z_{eff}, gives the 'correct' square of the overlap. By adopting a certain smooth function $\rho(x)$ and choosing the appropriate size of the nucleus, Z_{eff} is calculalted by computer for some 30 nuclei between Be and U. A proper interpolation enables us to find Z_{eff}.

A further develoment is the introduction of the concept of forbiddenness. A Rayleigh expansion in (24)

$$e^{-ip_\nu \cdot r_i} = \sum_l A_l j_l (p_\nu r_i) Y_l^0 (\cos \theta)$$

$$\theta = \sphericalangle (p_\nu, v_i)$$

is convenient in two ways. First, the convergence of j_l is pretty good: the ratio of j_{l+2}/j_l at the nuclear surface is estimated to be of the order of 1/30, hence the squared matrix element is 10^{-3} smaller if l increases two. Second, since the nuclear state is characterized by the eigenvalues of the angular momentum, we can obtain the selection rules according to the tensorial rank of the operator O. This program was fully carried out by Morita and Fujii.[1]

The reduced matrix element is characterized by three numbers s,w,u and the additional symbol $+$, $-$, and p:

$$[s, w, u] \qquad [s, w, u, +] \qquad [s ,w, u, -] \qquad [s, w, u, p]$$

The physical meaning of s is the resultant spin of the leptons and takes the value 0 or 1. w is the (effective) orbital angular momentum of the neutrino, u is the tensorial rank of the operator O. s, w, u give the selection rules between J_a and J_b. The additional symbol $\pm$ comes from the derivative of the muon orbital. We recognize

$$[s, w, u, -] \approx [s, w, u]$$

but $[s, w ,u, +]$ may be much different from $[s, w, u]$ itself. The symbol p means to differentiate with respect to the proton wavefunction. Estimating the relative magnitude of these reduced matrix elements, we can classify them according to forbiddenness as in Table 3. The usefulness of the forbiddenness is recently confirmed in O^{16} experiment.[22]

Table 3

Selection Rules for Muon Capture in Nuclei

Forbiddenness	Parity charge	$\Delta J = \|J_a - J_b\|$	The number of reduced matrix element
Allowed	No(+)	0,1	9
1st-Forbidden	Yes(−)	0,1,2	17
2nd-Forbidden	No	2,3	14
3rd-Forbidden	Yes	3,4	14
nth-Forbidden	$(-1)^n$	$n, n+1$	14

5. EXPERIMENTAL PARTIAL CAPTURE RATE

The experiments on the partial capture rate between the ground states (except O^{16}, in which the partial rate to several bound states are also measured) are summarized in Figs. 4, 5, and 6. For the sake of reference, the calculations along the same line are summarized in Figs. 7 and 8 in which, however, no experiment has been performed so far.

At present, the interest is concentrated to obtain a reliable estimate of the induced pseudoscalar coupling constant $C_P = (m_\mu/m)g_2$ in the elementary process. It is, of course, subject to the errors due to the nonrelativistic approximation in the elementary process on the one hand, and to the impulse approximation and the nuclear wavefunction or nuclear model on the other. The values of $-C_P/C_A = (m_\mu/m)\cdot(g_2/g_1)$, deduced from these experiments, are summarized in Table 4.

Table 4

The Induced Pseudoscalar Coupling from the Partial Capture Rate

Reaction	$-\frac{C_p}{C_A} = \frac{m_\mu}{m}\frac{g_2}{g_1}$
$\mu^- + He^3 \longrightarrow H^3 + \nu$	$8 \sim 10$
$\mu^- + C^{12} \longrightarrow B^{12} + \nu$	8 ± 2
$\mu^- + O^{16} \longrightarrow N^{16} + \nu$	16

As we mentioned in Section 3, the reaction $\mu + He^3 \rightarrow H^3 + \nu$ is particularly important, because of the experimental accuracy and

1. $\mu^- + C^{12} \longrightarrow B^{12} + \nu$

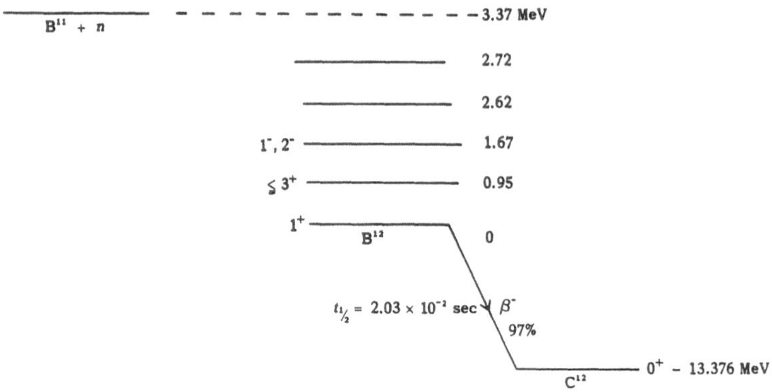

Fig. 4

Experiment

Author	Detector	Rate, 10^3 sec^{-1}
Burgman et al. [23]	counter	9.12 ± 0.45
*Argo et al. [24]	counter	9.05 ± 0.95
Fetkovich et al. [25]	bubble chamber	6.8 ± 1.1
Love et al. [26]	counter	6.8 ± 1.5
Reynolds et al. [27]	filament scintilation counter	6.7 ± 0.9
Maier et al. [28]	counter	6.31 ± 0.24
*Godfrey [29]	counter	5.9 ± 1.5
Bloch [30]	counter	5.8 ± 1.3

* Cosmic ray as muon source

Theory

Author	Nuclear model	Nuclear radius, 10^{-13}cm	Rate, 10^3* sec^{-1}
Fujii-Primakoff [11]		2.52	7.39
Wolfenstein [31]			6.9
Flamand-Ford [32]	intermediate	1.65	7.26
Morita-Fujii [33]		2.52	6.69
Ruel-Breman [34]			6.85

* Published values corrected taking the finite nuclear size correction −6%.

the theoretical confidence with relativistic structure parameters (form factors). The Shapiro ratio in the reaction $\mu + O^{16} \rightarrow N^{16} + \nu$ depends on the induced pseudoscalar coupling sensitively, since it is directly responsible to $0^+ \rightarrow 0^-$ transition.

The conclusion from the nuclear capture is not much superior to the capture in hydrogen in degree of confidence, but our original picture of the elementary process seems to be consistent with the experiment.

2. $\mu^- + He^3 \longrightarrow H^3 + \nu$

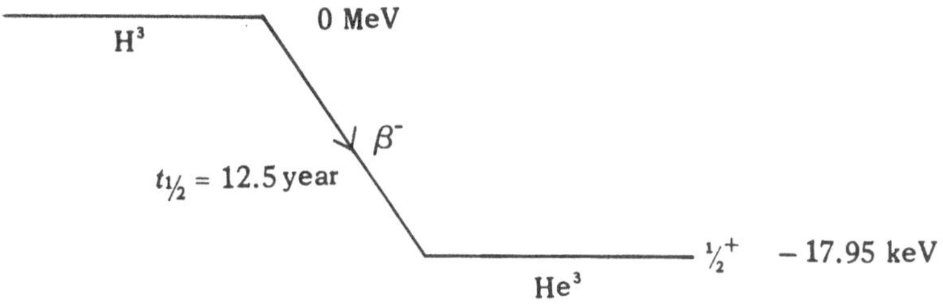

Fig. 5

Experiment

Author	Detector	Rate, 10^3 sec^{-1}
Auerbach et at. [35]	counter	1.52 ± 0.05
Falomkin et at. [36]	diffusion chamber	1.41 ± 0.14
Edelstein et al. [37]	counter	1.44 ± 0.09

Theory

Author	Wavefunction	Nuclear radius, 10^{-13} cm	Rate, 10^3 sec^{-1}
Fujii-Primakoff [19]	Irving	1.78	1.46
Fujii [38]	Kikuta-Morita-Yamada	1.50	1.66
Werntz [39]	Kikuta-Morita-Yamada	1.56	1.56
Duck [40]	Elliot-Flowers	1.0	1.25
Fujii-Yamaguchi [15]	Relativistic Form Factors	1.97	1.528
Drechster-Stech [16]	Relativistic Form Factors		~1500

3. $\mu^- + O^{16} \longrightarrow N^{16} + \nu$

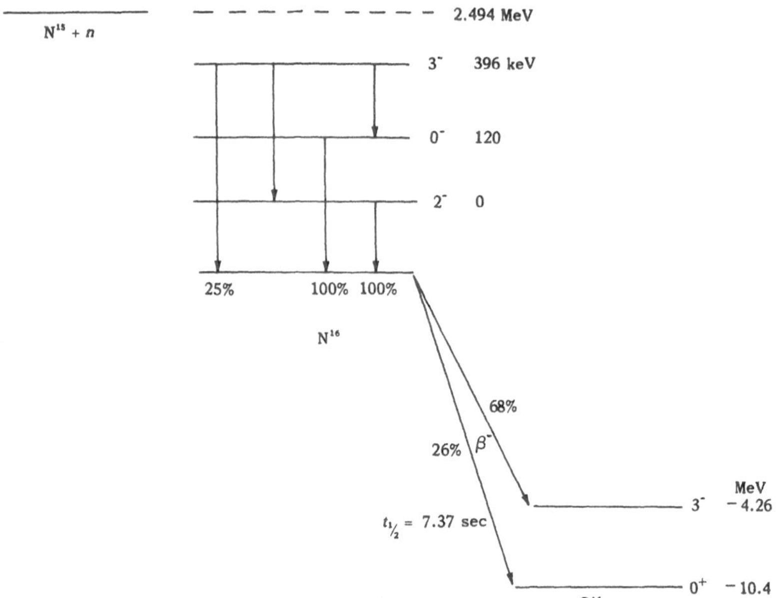

Fig. 6

Experiment

Author	Final State	Rate, 10^{-3} sec^{-1}
Cohen et al. [22]	2^-	6.76 ± 0.71
	0^-	0.76 ± 0.11
	3^-	Unobserved
	1^-	1.73 ± 0.10
	Ratio $0^-/1^-$	0.38 ± 0.07
Segre et al. [41]	Ratio $0^-/1^-$	1.23 ± 0.12

Theory

Author	Nuclear model	Nuclear radius, 10^{-13}	Rate, 10^{-13} sec^{-1}	
Shapiro-Blockimtsev [42]			$0^-/1^-$	0.12
Duck [40]		1.50	2^-	14.0
			$0^-/1^-$	0.33
	Elliott-Flowers	1.50 $\longrightarrow$	2^-	14.2
			$0^-/1^-$	0.66

Theory (*Continued*)

Author	Nuclear model	Nuclear radius, 10^{-13}	Rate, 10^{-13} sec^{-1}	
Ericson-Sens-Rood	[43] Elliott-Flowers	1.56	2^-	11.4
			0^-	1.15
			1^-	1.84
			$0^-/1^-$	0.63

* * *

4. $\mu^- + Li^6 \longrightarrow He^6 + \nu$

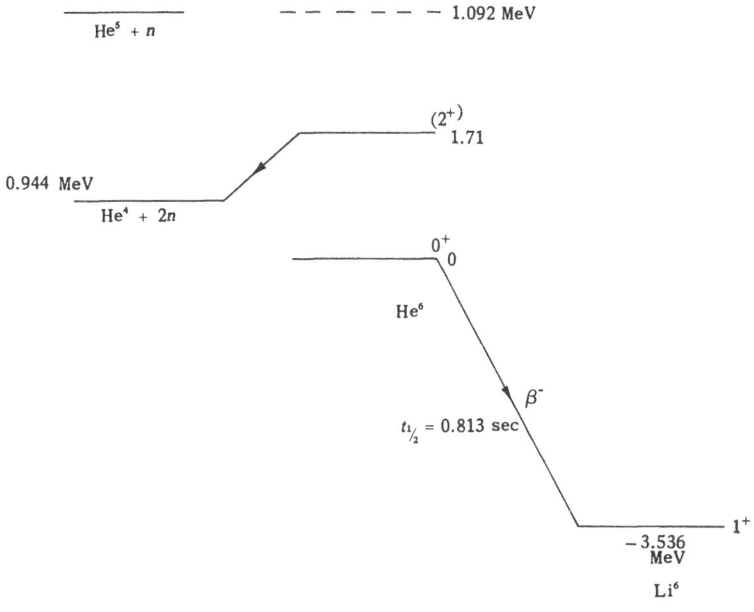

Fig. 7

Theory

Author	Nuclear model	Shell radius, 10^{-13}	Rate, 10^{-3} sec^{-1}
Fujii-Primakoff [19]	LS	2.40	1.79
Uberall [44]	LS	4.1	0.409
Delorme [45]	LS	4.1	0.702

5. $\mu^- + F^{19} \longrightarrow O^{19} + \nu$

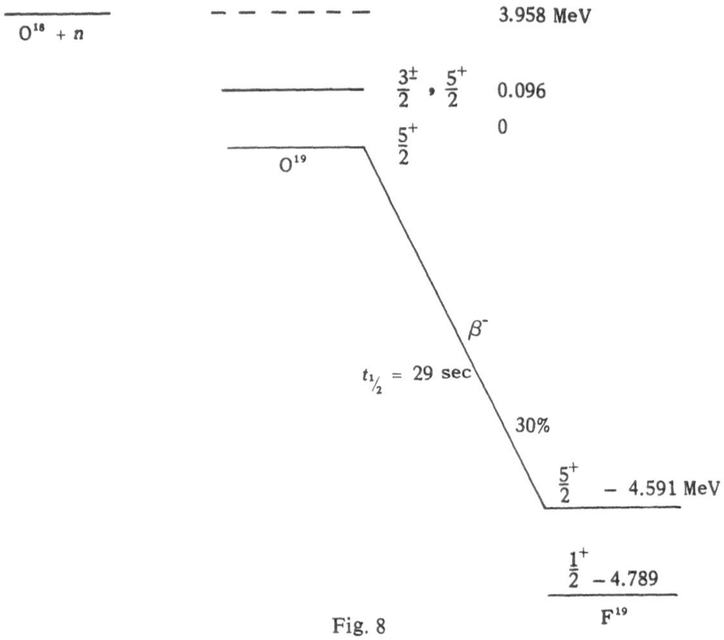

Fig. 8

Theory

Author	Nuclear model	Rate, 10^3 sec^{-1}
Duck [40]	Elliott-Flowers	1.35

SUMMARY AND REMARKS

The theory of the muon capture in nuclei rests on the following picture. For the elementary process

$$\mu^- + p \rightarrow n + \nu$$

we assume:

1) Lorentz invariance,
2) Time reversal invariance,
3) Definite *G*-parity,

$$GV_\lambda G^{-1} = V_\lambda \qquad GP_\lambda G^{-1} = -P_\lambda$$

4) c. v. c. hypothesis,

$$\partial_\lambda V_\lambda = 0$$

5) a. c. a. c. hypothesis,

$$\partial_\lambda P_\lambda \propto \phi_\pi$$

and for the nuclear process

$$\mu^- + (A, Z) \rightarrow (A, Z-1) + \nu$$

we introduce further

6) impulse approximation,

7) first order nonrelativistic approximation,

and rest our theoretical prediction upon a certain

8) nuclear model or nucleon wavefunction.

We are mainly interested in whether the dynamical principles 3), 4), and 5) work or not. The "nuclear physics" approximation 6), 7), and 8) may well be improved in future.

The conclusion is that our picture seems to be consistent with the experiment in general, although the errors due to the molecular or nuclear parameters do not allow a definite statement.

If the picture of the elementary process happens to be thoroughly confirmed, then we can turn the problem around and make use of the muon capture reaction in the study of the nuclear structure and the test of the nuclear wavefunction, etc. The intense muon beam may become available soon in pion factories, so that the muon capture reaction will become a useful tool as beta decay.

REFERENCES

1 M. Morita and A. Fujii, *Phys. Rev.* **118**: 606 (1960).

2 J. B. Adams, *Phys. Rev.* **126**: 1567 (1962).

3 E. Bleser, L. Lederman, J. Rosen, J. Rothberg, and E. Zavattini, *Phys. Rev. Letters* **8**: 128 (1962).

4 G. Conforts, S. Focardi, C. Rubbia, and E. Zavattini, *Phys. Rev. Letters* **9**: 432 (1962); *ibid.* **9**: 525 E (1962).

5 S. S. Gershtein, *Soviet Physics JETP* **7**: 318 (1958), *ibid.* **7**: 685 (1958).

6 S. Cohen, D. L. Judd, and R. J. Riddell, *Phys. Rev.* **119**: 397 (1963).

7 Ya. B. Zeldevich and S. S. Gershtein, *Soviet Physics Uspekhi* **3**: 593 (1961).

8 V. B. Belyaev, S. S. Gershtein, B. N. Zakharev, and S. P. Zemnev, *Soviet Physics JETP* **10**: 1171 (1961)

9 Ya. B. Zeldovich and S. S. Gershtein, *Soviet Physics JETP* **8**: 451 (1959).

10 M. Shimzu, Y. Mizuno, and T. Izuyama, *Progr. Theoret. Phys.* **20**: 77 (1958); **21**: 279 (1959).

11 S. Weinberg, *Phys. Rev. Letters* **4**: 575 (1960).

[12] A. Fujii, *Nuovo Cimento* **27**: 1025 (1963).
[13] C. Rubbia, *Proceedings of the International Conference on Fundamental Aspects of Weak Interactions Held at Brookhaven National Laboratory* (1963), p. 278.
[14] A. Halpern, *Phys. Rev.* **135A**: 34 (1964).
[15] A. Fujii and Y. Yamaguchi, *Prog. Theoret. Phys.* **31**: 107 (1964).
[16] W. Drechsler and B. Stech, *Zb. für Physik* **178**: 1 (1964).
[17] H. Collard *et al.*, *Phys. Rev. Letters* **11**. 132 (1963).
[18] L. B. Auerbach *et al.*, *Phys. Rev. Letters* **11**: 23 (1963).
[19] A. Fujii and H. Primakoff, *Nuovo Cimento* **12**: 327 (1959).
[20] L. L. Foldy and S. A. Wouthuysen, *Phys. Rev.* **78**: 29 (1950).
[21] J. C. Sens, *Phys. Rev.* **113**: 679 (1959).
[22] R. C. Cohen *et al.*, *Phys Rev. Letters* **11**: 134 (1963).
[23] J. O. Burgman, J. Fischer, B. Leontic, A. Lundby, R. Meunier, J. P. Stroot, and J. D. Teja, *Phys. Rev. Letters* **1**: 469 (1958).
[24] H. V. Argo, F. B. Harrison, H. W. Kruse, and A. D. McGuire, *Phys. Rev.* **114**: 626 (1959).
[25] J. G. Fetkovich, T. H. Fields, and R. L. McIlwain, *Phys. Rev.* **118**: 319 (1959).
[26] W. A. Love, S. Marder, I. Nadelhaft, R. T. Siegel, and A. E. Taylor, *Bull. Ann. Phys. Soc.* **4**: 81 (1959).
[27] G. T. Reynolds, D. B. Scarl, R. A. Swamm, J. R. Waters, and R. A. Zdanis, *Phys. Rev.* **129**: 1790 (1963).
[28] E. J. Maier, B. L. Bloch, R. M. Edelstein, and R. T. Siegel, *Phys. Rev. Letters* **6**: 417 (1961).
[29] T. N. K. Godfrey, *Phys. Rev.* **92**: 512 (1953).
[30] B. L. Bloch, Thesis, Carnegie Institute of Technology (1960).
[31] L. Wolfenstein, *Nuovo Cimento* **13**: 319 (1959).
[32] G. Flammand and K. W. Ford, *Phys. Rev.* **116**: 1951 (1959).
[33] M. Morita and A. Fujii, *Nuovo Cimento* **15**: 850 (1960).
[34] M. Ruel and J. B. Brennan, *Phys. Rev.* **129**: 366 (1963).
[35] L. B. Auerbach, R. J. Esterbing, R. E. Hill, D. A. Jenkino, J. T. Lach, and N. H. Lipman, *Phys. Rev. Letters* **11**: 28 (1963).
[36] I. V. Falomkin, A. I. Filippov, M. M. Kulyukin, B. Pontecorvo, Yu. A. Scherbokov, R. M. Sulyaev, V. M. Tsuppo-Sitinkov, and O. A. Zaimidoronga, *Phys. Letters*, **3**: 229 (1963).
[37] R. M. Edelstein, D. Clay, J. W. Keuffel, and K. L. Wagner, Jr., *Proceedings of the International Conference on the Fundamental Aspects of the Weak Interaction* (1963), p. 303
[38] A. Fujii, *Phys. Rev.* **118**: 870 (1960).
[39] C. Werntz, *Nucl. Phys.* 16: 59 (1960).
[40] I. Duck, *Nucl. Phys.* **35**: 27 (1962).
[41] E. Segre, Private communication.
[42] J. Shapiro and L. Blockintsev, *Soviet Physics JETP* **12**: 775 (1961).
[43] T. Ericson, J. Sens, and H. P. C. Rood, CERN Preprint, Sept. 1963.
[44] H. Uberall, *Phys. Rev.* **116**: 218 (1959).
[45] J. Delorme, *Nuovo Cimento* **32**: 1360 (1964).

Quantum Gauge Transformations

JERZY LUKIERSKI

UNIVERSITY OF WROCLAW
Wroclaw, Poland

1. INTRODUCTION

The gauge transformations of the first and the second kind, as used in particle physics, are taken from classical field theory. In order to explain them, let us consider the simplest case: classical electrodynamics. We know that the gauge transformations of the second kind

$$A'_{\mu}(x) = A_{\mu}(x) - \partial_{\mu}\alpha(x) \tag{1}$$

do not change the classical Maxwell equations which describe the physical content of the theory. From the form of (1) it follows that the change of the value of $A_{\mu,}{}^{\mu}(x)$ does not affect the observables, and the arbitrariness of $\alpha(x)$ in (1) corresponds to the freedom of choice of the scalar component $A_{\mu,}{}^{\mu}(x)$.

Now let us go onto quantum field theory. All c-number wave-functions become operators. In order to express the arbitrariness of the operator $\hat{A}_{\mu,}{}^{\mu}(x)$, we have to introduce an arbitrary gauge operator $\hat{\alpha}(x)$. The result is a special case of a more general rule: that *the arbitrary functional parameters $\alpha_{\iota}(x)$ determining in classical theory the gauge transformations of the second kind become the arbitrary gauge operators $\hat{\alpha}_{\iota}(x)$ in quantum field theory.**

*This is true only for the Heisenberg picture. In the interaction picture, where Green's functions are constructed from the free two-point propagators via perturbation theory, we can only introduce, in a natural way, the simplest operator gauge transformation, changing arbitrarily the nonphysical part of the free two-point propagator. These transformations have been studied for spinor electrodynamics in the most complete way by Okubo.[1]

Now we shall describe briefly how one can parameterize the arbitrariness of the operator. It is known that for the algebra A of the smeared-out operators $\hat{\alpha}_i[f]$ every linear form on A, which is a functional on A, determines the representation of A.† If we assume that this functional is analytic for $f \equiv 0$, the arbitrariness of choice of $\hat{\alpha}_i(x)$ will correspond to the arbitrariness of the coefficients in the Volterra expansion, which form the set of quantum gauges $M^{(n)}(x_1 \cdots x_n)(n = 1, 2, 3 \cdots)$. Describing the whole system by the set of time-ordered functions $\tau^{(n)}$, we shall obtain the nonphysical infinite-dimensional quantum gauge transformations

$$\delta\tau^{(n)} = \alpha^{(n)}_{(n';k)}\, \delta M^{(k)} \tau^{(n')} \tag{2}$$

where the set $\delta M^{(n)}(x_1 \cdots x_n)$ is chosen to have the symmetries of the time-ordered product, and measures infinitesimal change of $\hat{\alpha}_i$; the infinite-dimensional matrices $(\alpha^{(k)})_{nn'} = \alpha^{(n)}_{(n';k)}$ are the generators of quantum gauge transformations of $(k+1)$-th kind in the space of τ functions.

The reason for the existence of the transformations (2) is because the space, on which the gauge-dependent field operators act, is too large, and all the information about the observables (physical quantities) can be extracted from the subspace (sometimes called the subspace of physical solutions). The theory in the remaining, nonphysical part of the space can, in principle, be assumed arbitrarily, and the operator gauges display the variety of all possible theories, among which the differences are physically nonrelevant. The formula (2) describes the connection between two such theories, and represents the necessary and sufficient condition.

An example of a quantum gauge was investigated first in electrodynamics by Landau and Khalatnikov.[3] They introduced the two-point photon propagator with an arbitrary longitudinal part. The first authors who introduced the method of generating functional to study the simplest quantum gauges in electrodynamics were Zumino[4] and Bialynicki-Birula.[5]

The aim of studying the quantum gauges is to split all the calculational results, especially containing divergences, into terms depending on the choice of the gauge operator (or quantum gauges)

†If the linear form is positive definite, we have the so-called Gelfand construction[2] and we obtain a positive definite Hilbert space. Because $\hat{\alpha}_i$ are not physical, the space defined by the assumed functional can be more general. This, of course, will be reflected in the lack of the positive-definiteness condition.

and those independent of the above choice. For example, renormalization constants and, in general, the asymptotic behaviors, will depend on the choice of quantum gauges.

It is well-known that gauge-invariance leads to the vanishing of the bare mass. It is hoped that these studies can have some meaning for the Wightman formulation of the theory of mass-less fields.

2. Quantum Gauge Transformations in Electrodynamics

We shall at first introduce the spinor electrodynamics with the electromagnetic field A_μ having only spin one degrees of freedom. This choice of A_μ is known in quantum theory as the Landau gauge.

Let us introduce the following Lagrangian:

$$\mathscr{L} = -\tfrac{1}{4} F_{\mu\nu} F^{\mu\nu} - \lambda \partial^\mu A_\mu + e\bar{\psi}\gamma^\mu \psi A_\mu + \mathscr{L}^0_\psi \tag{3}$$

where λ is a functional Lagrange multiplier.

We can quantize the system described by (3), considering $\lambda(x)$ as an independent field variable. The constraints disappear and we have

$$\begin{aligned}
&\text{(a)} \qquad [F_{i0}(\mathbf{x}, t), A_j(\mathbf{x}', t')]_{t=t'} = i\delta_{ij}\delta(\mathbf{x} - \mathbf{x}') \\
&\text{(b)} \qquad [\lambda(\mathbf{x}, t), A_0(\mathbf{x}, t')]_{t=t'} = i\delta(\mathbf{x} - \mathbf{x}') \\
&\text{(c)} \qquad \{\psi^*_\alpha(\mathbf{x}, t), \psi_\beta(\mathbf{x}', t')\}_{t=t'} = \delta(\mathbf{x} - \mathbf{x}')
\end{aligned} \tag{4}$$

and all other commutators vanish.

The field equations look as follows:

$$\begin{aligned}
&\text{(a)} \qquad \partial^\mu F_{\mu\nu} = -\partial_\nu \lambda \\
&\text{(b)} \qquad \{\gamma^\mu(\partial_\mu - ieA_\mu) + m\}\psi = 0 \\
&\text{(c)} \qquad \partial^\mu A_\mu = 0
\end{aligned} \tag{5}$$

where we would like to point out that the vanishing of the longitudinal component is not a subsidiary condition but one of the field equations.

The equations (5) and the commutation relations (4) enable us to define the generating functional

$$Z[J, \eta, \bar{\eta}, \Lambda] = <0|T \exp i\left\{\int dx(\bar{\eta}\psi + \bar{\psi}\eta + J_\mu A^\mu + \Lambda\lambda)\right\}|0> \tag{6}$$

by means of the following functional differential equations:

$$\text{a)}\quad \left\{\partial^\lambda\left(\partial_\lambda \frac{1}{i}\frac{\delta}{\delta J^\mu} - \partial_\mu \frac{1}{i}\frac{\delta}{\delta J^\lambda}\right) - ie\frac{1}{i}\frac{\delta}{\delta\eta}\gamma_\mu\frac{1}{i}\frac{\delta}{\delta\bar\eta} + \partial_\mu\frac{1}{i}\frac{\delta}{\delta\Lambda}\right\} Z = J_\mu Z$$

$$\text{b)}\quad \left\{\gamma^\mu\left(\partial_\mu - ie\frac{1}{i}\frac{\delta}{\delta J^\mu}\right) + m\right\}\frac{1}{i}\frac{\delta}{\delta\bar\eta} Z = \eta Z$$

$$\text{c)}\quad \frac{1}{i}\frac{\delta}{\delta\eta}\left\{-\gamma^\mu\left(\partial_\mu + ie\frac{1}{i}\frac{\delta}{\delta J^\mu}\right) + m\right\} Z = \bar\eta Z$$

$$\text{d)}\quad \partial^\mu \frac{1}{i}\frac{\delta}{\delta J^\mu} Z = \Lambda Z \tag{7}$$

Multiplying (7a) by ∂^μ and using (7 b–c) we get

$$\Box\frac{1}{i}\frac{\delta}{\delta\Lambda} Z = \left(\partial^\mu J_\mu + e\eta\frac{\delta}{\delta\eta} - e\bar\eta\frac{\delta}{\delta\bar\eta}\right) Z \tag{8}$$

The equation (8) describes the dependence of Z upon Λ. The c-number field Λ is an external field influencing λ. *Because λ is coupled only to $A_{\mu,}{}^{\mu}$ and changes of $A_{\mu,}{}^{\mu}$ are not physical, we can conclude that λ measures the gauge functions.*

Before further studies of the gauge transformations, we shall make the following substitution in equations (6) to (8)

$$\Lambda(x) \to \tilde\Lambda(x) = \int D(x - x')\Lambda(x')\,dx' \tag{9}$$

where $\Box\, D(x - x') = \delta(x - x')$.

This corresponds to the following replacement of the subsidiary condition:[5]

$$A_{\mu,}{}^{\mu}(x) \to R[A; x] = \int a_\mu(x - x')A^\mu(x')\,dx' \tag{10}$$

where

$$\partial^x_\mu a^\mu(x - x') = \delta(x - x') \tag{10a}$$

Because we get after performing the gauge transformation (1)

$$R[A'; x] = R[A; x] - \alpha(x) \tag{11}$$

we see that the new subsidiary condition R describes exactly the degree of freedom characterised by the gauge operator. We now shall rewrite (8) with the replacement (9).

$$\frac{1}{i}\frac{\delta}{\delta\tilde\Lambda(x)} Z = \left(\partial^\mu J_\mu(x) + e\eta(x)\frac{\delta}{\delta\eta(x)} - e\bar\eta(x)\frac{\delta}{\delta\bar\eta(x)}\right) Z \tag{12}$$

The equation (12) determines the response of the system to the external current $\tilde{\Lambda}(x)$. The quantum gauge transformations measure the higher responses of the systems. We introduce

$$Z = Z[J, \eta, \bar{\eta}; \{M\}] \tag{13a}$$

where $\{M\} \equiv (\tilde{\Lambda} = M^{(1)}, M^{(2)}, M^{(3)} \cdots)$ and

$$\frac{1}{i}\frac{\delta}{\delta M^{(n)}(x_1\cdots x_n)} Z = \frac{1}{n!}\prod_{j=1}^{n}\frac{1}{i}\frac{\delta}{\delta\tilde{\Lambda}(x_j)} Z \tag{13b}$$

We see from (12) and (13b) that all $M^{(n)}$ are symmetric with respect to any interchange of the arguments.

The generating functional (13a) can be introduced if we add to (3) the following free Lagrangian of the field λ

$$\int \mathscr{L}\, dx \to \int \mathscr{L}\, dx + \mathscr{L}_0[\lambda] \tag{14}$$

where

$$\mathscr{L}_0[\lambda] = \sum_{n=2}^{\infty}\frac{1}{n!}\int dx_1 \cdots \int dx_n M^{(n)}(x_1\cdots x_n)\lambda(x_1)\cdots\lambda(x_n) \tag{15}$$

We see that $M^{(n)}(x_1\cdots x_n)$ are the coefficients of the Volterra expansion of the most general free Lagrangian for the Lagrange multiplier field. In this formulation the arbitrariness of the gauge operator is expressed by the freedom of choice of the Lagrange multiplier λ, which is coupled to R. The formula (13b) follows from (15) immediately, if we use the representation of the generating functional (13a) by the all-path Feynman integral.[6]

Now we shall calculate the transformation properties of the propagators. The photon propagators are defined as follows:

$$G_{\mu_1\cdots\mu_n}(x_1\cdots x_n) = \frac{\delta}{\delta J_{\mu_1}(x_1)}\cdots\frac{\delta}{\delta J_{\mu_n}(x_n)}\frac{1}{i}\ln Z\Big|_{\substack{J=0\\ \eta=\bar{\eta}=0}} \tag{16}$$

Because

$$\left[\frac{\delta}{\delta J_\mu}, \frac{\delta}{\delta M^{(n)}}\right] = 0 \qquad n = 1, 2, 3\ldots \tag{17}$$

we have

$$\frac{\delta G_{\mu_1\cdots\mu_n}(x_1\cdots x_n)}{\delta M^{(m)}(y_1\cdots y_n)} = \frac{\delta}{\delta J_{\mu_1}(x_1)}\cdots\frac{\delta}{\delta J_{\mu_n}(x_n)}\frac{1}{i}\frac{\delta}{\delta M^{(m)}(y_1\cdots y_n)}\ln Z\Big|_{\substack{J=0\\ \eta=\bar{\eta}=0}} \tag{18}$$

Using (12) and (13b), we get the following quantum gauge

transformations for (16):

$$\delta_{M(n)} G_{\mu_1\cdots\mu_n}(x_1\cdots x_n) = \begin{cases} 0 & m \neq n \\ \partial^{x_1}_{\mu_1}\cdots\partial^{x_n}_{\mu_n}\delta M^{(n)}(x_1\cdots x_n) & m = n \end{cases} \tag{19}$$

We see that each quantum gauge affects exactly one propagator. The gauge transformations of second kind $(m = 1)$ change the average field G_μ, the gauge transformations of third kind $(m = 2)$ the two-point propagator, and, in general, the gauge transformation of the $N + 1$-th kind $(m = N)$— the N-point propagator. The formulas (19) can be directly integrated, and this amounts to the replacement of δM by M. For average field G_μ, we get (1) with $\alpha \equiv \tilde{\Lambda} = M^{(1)}$, and choosing

$$M^{(2)} = a\frac{1}{\square}\cdot\frac{1}{\square - i\varepsilon}$$

we obtain for $a = -1$ the Feynman form of the photon propagator and for $a = -3$ the gauge used by Fried and Yennie,[7] compensating the infrared divergences.*

In order to calculate the quantum gauge transformations for the electron propagator

$$G(x, y) = \frac{1}{Z}\frac{1}{i}\frac{\delta}{\delta\eta(y)}\frac{\delta}{\delta\bar{\eta}(x)} Z\Big|_{\substack{J=0\\ \eta=\bar{\eta}=0}} \tag{20}$$

we have to differentiate (12) or, correspondingly, (13b) with respect to η and $\bar{\eta}$. We get for $m = 1$, observing that $Z[\delta\tilde{\Lambda}]\Big|_{\substack{J=0\\ \eta=\bar{\eta}=0}} = Z[0]\Big|_{\substack{J=0\\ \eta=\bar{\eta}=0}}$

$$\delta_\Lambda G(x, y) = ie[\delta\tilde{\Lambda}(x) - \delta\tilde{\Lambda}(y)]\, G(x, y) \tag{21a}$$

which gives after integration

$$G'(x, y) = \exp ie[\tilde{\Lambda}(x) - \tilde{\Lambda}(y)]\cdot G(x, y) \tag{21b}$$

For $m = 2$, since $Z[\delta M]\Big|_{\substack{J=0\\ \eta=\bar{\eta}=0}} = Z[0]\Big|_{\substack{J=0\\ \eta=\bar{\eta}=0}}$

we have

$$\delta_{M(2)} G(x, y) = ie^2[\delta M(x - y) - \delta M(0)]\, G(x, y) \tag{22}$$

where we have assumed that $\delta M(x, y) = \delta M(x - y) = \delta M(y - x)$.

* The choice $a = 0$, we recall, corresponds to the Landau gauge.

Integrating (22) we get

$$G'(x, y) = \exp ie^2[M(x-y) - M(0)] \cdot G(x, y) \tag{23}$$

Inserting in (23) the large separation of coordinates we obtain

$$Z_2' = \exp\{-ie^2 M(0)\} Z_2 \tag{24}$$

where Z_2 is a wave renormalization constant. The formula (24) was obtained first by Johnson and Zumino.[8]

The electron propagator also depends on the higher quantum gauge functions. For example, if we take the quantum gauge transformations of the fifth kind, characterized by the gauge function $M^{(4)}(x_1 \cdots x_4)$ (symmetric in all indices), then we obtain the formula

$$G'(x, y) = \exp ie^4 M^{(4)}(x, y) \cdot G(x, y) \tag{25}$$

where

$$\begin{aligned} M^{(4)}(x, y) = \tfrac{1}{4}\{&M^{(4)}(x, x, y, y) - \tfrac{1}{2}M^{(4)}(x, y, y, y) \\ &- \tfrac{1}{2}M^{(4)}(y, x, x, x) - \tfrac{1}{6}M^{(4)}(x, x, x, x) + \tfrac{1}{6}M^{(4)}(y, y, ,y\, y)\} \end{aligned} \tag{25a}$$

From (25a) we see that $M^{(4)}$ can be represented by a properly chosen gauge transformation of the third kind. There is true a general rule that for the N-th point Green's function all possible quantum gauge transformations can be expressed by the first $N+1$ quantum gauges.

We can calculate further, for example, the transformation properties of the four-point Green functions $G(x, y, z, u)$ describing scattering of two electrons, under the quantum gauge transformations of the fifth kind. The formula is lengthy and exhibits an exponential dependence of the phase-factor type, as in (25), with e^4 in the exponential.

Finally, we calculate the transformation properties of the vertex functions, with photon and electron external lines. The method is to calculate (23) for $J \neq 0$ and then differentiate with respect to J_μ.

For the three-point vertex function we get[4]:

$$\begin{aligned} C_\mu'(x, y; z) &= \frac{\delta}{\delta J^\mu(z)} G'(x, y; J)|_{J=0} \\ &= \exp ie^2[M(x-y) - M(0)] \cdot \Big\{C_\mu(x, y; z) \\ &\qquad - ieG(x, y)\frac{\partial}{\partial z_\mu}[M(x-z) - M(y-z)]\Big\} \end{aligned} \tag{26}$$

3. QUANTUM GAUGE TRANSFORMATIONS FOR NON-ABELIAN GAUGE VECTOR FIELDS.

The first complication which one meets when considering the non-linear gauge-invariant theory of Yang-Mills type[9] is the problem of the proper subsidiary condition. In order to identify the degrees of freedom described by the gauge-functions $\alpha_i (i = 1 \cdots n)$ or by the subsidiary conditions R_i, we have to introduce R_i by an equation which is the generalization of (11).

$$\frac{\delta R_i[B; x]}{\delta \alpha_j(y)} = -\delta_{ij}\delta(x-y) \tag{27}$$

Because the infinitesimal gauge transformation for non-Abelian gauge field looks as follows:[10]

$$\delta_\alpha B_{\mu;i} = -\partial_{\mu;ij}[B]\,\delta\alpha_j \tag{28}$$

where

$$\frac{\delta_\alpha B_{\mu;i}}{\delta \alpha_j} = \partial_{\mu;ij}[B] = \partial_\mu \delta_{ij} - igt_{ikj}B_{\mu;k} \tag{29}$$

and t_{ijk} are the structure constants of the classical Lie gauge groups, the equation (27) is equivalent to

$$\frac{\delta R_i[B; x]}{\delta B_{\mu;k}(y)}\,\partial^y_{\mu;kj}[B] = -\delta_{ij}\delta(x-y) \tag{30}$$

Equation (30) is the generalization of (10a).

We shall write first the exact equations for the generating functional Z, assuming Lorentz gauge. Because even the free non-Abelian vector gauge field is charged, we shall obtain nontrivial results without adding the fermion sources. The Lagrangian describing non-Abelian vector gauge fields with vanishing longitudinal components looks as follows[11]:

$$\mathscr{L} = -\tfrac{1}{4}F_{\mu\nu;i}F^{\mu\nu;i} - \lambda_i \partial^\mu B_{\mu;i} \tag{31}$$

where λ_i are the Lagrange multipliers, and

$$F_{\mu\nu;i} = \partial_\mu B_{\nu;i} - \partial_\nu B_{\mu;i} - igt_{ijk}B_{\mu;j}\circ B_{\nu;k} \tag{32}$$

and $x \circ y = \frac{1}{2}(xy + yx)$.

From (31) we get the following equal time canonical commutation relations:

$$[F_{ro;i}(\mathbf{x}, t), B_{s;j}(\mathbf{x}', t')]_{t=t'} = i\delta_{rs}\delta_{ij}\delta(\mathbf{x}-\mathbf{x}')$$
$$[\lambda_i(\mathbf{x}, t), B_{o;j}(\mathbf{x}', t')]_{t=t'} = i\delta_{ij}\delta(\mathbf{x}-\mathbf{x}') \tag{33}$$

and other commutators vanish.

The field equations are

$$D_{\nu;i}[B] = \partial^\mu F_{\mu\nu;i} = -\partial_\nu \lambda_i \tag{34}$$
$$\partial^\mu B_{\mu;i} = 0$$

We can introduce the generating functional

$$Z[\underline{J}; \underline{\Lambda}] = <0|T \exp i \left\{\int dx (B_{\mu;i} J^{\mu;}{}_i + \lambda_i \Lambda_i)\right\}|0> \tag{35}$$

which satisfies the following functional differential equations.

a) $$D_{\mu;i}\left[\frac{1}{i}\frac{\delta}{\delta \underline{J}}\right] Z[\underline{J}, \underline{\Lambda}] = \left(J_{\mu;i} - \partial_\mu \frac{1}{i}\frac{\delta}{\delta \Lambda_i}\right) Z[\underline{J}, \underline{\Lambda}]$$

b) $$\partial^\mu \frac{1}{i}\frac{\delta}{\delta J_{\mu;i}} Z[\underline{J}, \underline{\Lambda}] = \Lambda_i Z[\underline{J}, \underline{\Lambda}] \tag{36}$$

Multiplying of (36a) by $\partial_{\mu,ji}[(1/i)(\delta/\delta \underline{J})]$ leads to the following equation which describes the dependence of (35) on Λ_i:

$$\left\{\partial^{\mu;}{}_{ij}\left[\frac{1}{i}\frac{\delta}{\delta \underline{J}}\right] \partial_\mu \frac{1}{i}\frac{\delta}{\delta \Lambda_j} - \partial^{\mu;}{}_{ij}\left[\frac{1}{i}\frac{\delta}{\delta \underline{J}}\right] J_{\mu;j}\right\} Z[\underline{J}, \underline{\Lambda}] = 0 \tag{37}$$

Now we shall introduce a new $\tilde{\Lambda}_j$ by means of the formula

$$\partial^{\mu;}{}_{ij}\left[\frac{1}{i}\frac{\delta}{\delta \underline{J}}\right] \partial_\mu \frac{1}{i}\frac{\delta}{\delta \Lambda_j} = \frac{1}{i}\frac{\delta}{\delta \tilde{\Lambda}_i} \tag{38}$$

which corresponds to the following change of $\tilde{\lambda}_i$ in (34)

$$\lambda_i \to \tilde{\lambda}_i = \partial^{\mu;}{}_{ij}[B]\, \partial_\mu \lambda_j \tag{39}$$

The Lagrange multipliers $\tilde{\lambda}_i$ correspond—it can be seen from (35) and (39)—to the choice (30) of the subsidiary condition. The equation (37) now has the form

$$\frac{1}{i}\frac{\delta}{\delta \tilde{\Lambda}_i} Z[\underline{J}, \underline{\tilde{\Lambda}}] = \partial^{\mu;}{}_{ij}\left[\frac{1}{i}\frac{\delta}{\delta \underline{J}}\right] J_{\mu;j}\, Z[\underline{J}, \underline{\tilde{\Lambda}}]$$
$$= \left(\partial^\mu J_{\mu;i} - igt_{ijk} J_{\mu;k} \frac{1}{i}\frac{\delta}{\delta J_{\mu;j}}\right) Z[\underline{J}, \underline{\tilde{\Lambda}}] \tag{40}$$

and describes the gauge properties of the generating functional. Now, as in the case of electrodynamics, we shall introduce quantum gauge functions defined by higher responses of the system to the current $\tilde{\Lambda}_i$. It is easy to see, however, that the $\tilde{\Lambda}_i$-derivatives in (40) do not commute. This will imply complications if one wants to calculate from (40) finite gauge transformations. In order to define infinitesimal quantum gauge transformations [see formulas (13a, b)], we should introduce the *symmetrized* product of $\tilde{\Lambda}_i$-derivatives:

$$\frac{1}{i}\frac{\delta Z[\underline{J};\underline{\{M\}}]}{\delta M^{(n)}_{i_1\cdots i_n}(x_1\cdots x_n)}=\frac{1}{(n!)^2}\sum_{\substack{\text{perm}\\(\alpha_1\cdots\alpha_n)}}\prod_{j=1}^{n}\frac{1}{i^n}\frac{\delta}{\delta\tilde{\Lambda}_j(x_{\alpha j})}Z[\underline{J};\underline{\{M\}}] \tag{41}$$

where $\underline{\{M\}}\equiv(\tilde{\Lambda}_i=M^{(1)}_i, M^{(2)}_{ij}, M^{(3)}_{ijk},\cdots)$ describes the set of quantum gauge transformations. Now let us pass to the calculations of the transformation properties of the propagators. We have

$$G_{\mu_1\cdots\mu_n;i_1\cdots i_n}[x_1\cdots x_n;\underline{\{M\}}]=\prod_{j=1}^{n}\frac{\delta}{\delta J_{\mu_j;i_j}(x_j)}\cdot\frac{1}{i}\ln Z[\underline{J};\underline{\{M\}}]|_{\underline{J}=0} \tag{42}$$

Similarly, we can introduce the following correlation functions:*

$$\kappa_{i_1\cdots i_n}[x_1\cdots x_n;\underline{J};\underline{\{M\}}]=\frac{1}{n!}\sum_{\substack{\text{perm}\\(\alpha_1\cdots\alpha_n)}}\prod_{j=1}^{n}\frac{\delta}{\delta\tilde{\Lambda}i_{\alpha_j}(x_{\alpha_j})}\cdot\frac{1}{i}\ln Z[\underline{J};\underline{\{M\}}] \tag{43}$$

We have, for the gauge transformations of the second kind, the following formula:

$$\delta_\Lambda G_{\mu_1\cdots\mu_n;i_1\cdots i_n}(x_1\cdots x_n)=\prod_{j=1}^{n}\frac{\delta}{\delta J_{\mu_j;i_j}(x_j)}\int dx\kappa_i[x;\underline{J}]\,\delta\tilde{\Lambda}_i(x)\Big|_{\underline{J}=0} \tag{44}$$

and for the (quantum) gauge transformation of the third kind,

$$\delta_{M^{(2)}}G_{\mu_1\cdots\mu_n;i_1\cdots i_n}(x_1\cdots x_n)=\prod_{j=1}^{n}\frac{\delta}{\delta J_{\mu_j;i_j}(x_j)}$$
$$\times\tfrac{1}{2}\int dxdy(\kappa_i[x;\underline{J}]\,\kappa_j[y;\underline{J}]+\kappa_{ij}[x,y;\underline{J}])\,\delta M_{ij}(x,y)\Big|_{\underline{J}=0} \tag{45}$$

Because

$$\kappa_i[x;\underline{J}]=\partial^x_\mu J^{\mu;}{}_i(x)-igt_{ijk}G_{\mu;j}[x;\underline{J}]\,J^{\mu;}{}_k(x) \tag{46}$$

and

$$\kappa_{ij}[x,y;\underline{J}]=ig^2(t^{ij})_{mn}G_{\mu;n}[x;\underline{J}]\,J^{\mu;}{}_m(x)\,\delta(x-y)$$
$$+ig^2t_{lik}t_{sjt}G_{\mu\nu;ls}[x,y;\underline{J}]\,J^{\mu;}{}_k(x)\,J^{\nu;}{}_t(y) \tag{47}$$

---

* It is interesting to mention that $\kappa_{i_1\cdots i_n}\neq 0$ for $n\geq 2$ only if $g\neq 0$. In other words, electrodynamics behaves classically, with only "average field" $\kappa[x;\tilde{\Lambda}]$ different from zero.

where

$$(t^{ij})_{mn} = \tfrac{1}{2}\{t_{mik}t_{kjn} + t_{mjk}t_{ijk}\} \tag{47a}$$

we get the following formula for the gauge transformations of the second kind:

$$\delta_\Lambda G_{\mu;i}(x) = igt_{ijk}\,\delta\tilde{\Lambda}_k(x)\,G_{\mu;j}(x) - \partial_\mu\delta\tilde{\Lambda}_i(x) \tag{48}$$

which is identical with (28) if $\delta\alpha_i = \delta\tilde{\Lambda}_i$, and

$$\begin{aligned}\delta_\Lambda G_{\mu_1\cdots\mu_n;i_1\cdots i_n}(x_1\cdots x_n) &= ig\sum_{\alpha=1}^{n} t_{i_\alpha jk}\\ &\times G_{\mu_1\cdots\mu_n;i_1\cdots i_{\alpha-1}j\,i_{\alpha+1}\cdots i_n}(x_1\cdots x_n)\cdot\delta\tilde{\Lambda}_k(x_\alpha)\end{aligned} \tag{49}$$

for $n \geqslant 2$.

The quantum gauge transformations of the third kind [see (45)] change the first two Green functions (42) as follows:

$$\delta_{M^{(2)}} G_{\mu;k}(x) = i\frac{g^2}{2}\,\delta M_{ij}(0)\,(t^{ij})_{ks}\,G_{\mu;s}(x) \tag{50}$$

and

$$\begin{aligned}\delta_{M^{(2)}}(G_{\mu\nu;kl}(x_1,x_2)) &= \{\partial_\mu^{x_1}\delta_{ik} + igt_{kik'}G_{\mu;k'}(x_1)\}\\ &\times\{\partial_\nu^{x_2}\delta_{jl} + igt_{ljl'}G_{\nu;l'}(x_2)\}\,\delta M_{ij}(x_1-x_2) + ig^2\\ &\times t_{kik'}t_{ljl'}\delta M_{ij}(x_1-x_2)\,G_{\mu\nu;k'l'}(x_1,x_2) + i\frac{g^2}{2}\\ &\times(t^{ij})_{kk'}\delta M_{ij}(0)\,G_{\mu\nu;k'l}(x_1,x_2) + i\frac{g^2}{2}(t^{ij})_{ll'}\delta M_{ij}(0)\\ &\times G_{\mu\nu;kl'}(x_1,x_2)\end{aligned} \tag{51}$$

where we assumed $\delta M_{ij}(x-y) = \delta M_{ji}(y-x)$.

From (51) we can see that $\delta M_{ij}(0)$ determines the wave renormalization constants. If we assume that $\lim_{z\to\infty}\delta M_{ij}(z)\to 0\,(z = x-y)$, we have the following differential formula for the wave renormalization matrix $Z_{kl;k'l'}$:

$$\delta Z_{kl;k'l'} = i\frac{g^2}{2}\{(t^{ij})_{kk'}\delta_{ll'} + \delta_{kk'}(t^{ij})_{ll'}\}\cdot\delta M_{ij}(0) \tag{52}$$

where

$$G_{\mu\nu;kl} = (\delta_{kk'}\delta_{ll'} + \delta Z_{kl;k'l'})\,G_{\mu\nu;k'l'} \tag{53}$$

One can calculate also the transformation properties for higher order Green functions, but the results are rather lengthy.[12]

REFERENCES

[1] S. Okubo, *Nuovo Cimento* **15**: 949 (1960).
[2] M. A. Naimark, *Normed Rings*, Moscow (1956).
[3] L. D. Landau, and I. M. Khalatnikov, *J. Exptl. Theor. Phys.* (*USSR*). **29**: 89 (1955).
[4] B. Zumino, *J. Math. Phys.* **1**: 1 (1960).
[5] I. Bialynicki-Birula, *J. Math. Phys.* **3**: 1094 (1962).
[6] J. Lukierski, *Nuovo Cimento* **29**: 561 (1963).
[7] H. M. Fried and D. R. Yennie, *Phys. Rev.* **112**: 1391 (1958).
[8] K. Johnson and B. Zumino, *Phys. Rev. Letters* **3**: 351 (1963).
[9] R. Mills and C. N. Yang, *Phys. Rev.* **96**: 191 (1954).
[10] J. Schwinger, *Phys. Rev.* **125**: 1043 (1962); *Phys. Rev.* **127**: 324 (1962).
[11] J. Schwinger, *Phys. Rev.* **130**: 402 (1963).
[12] J. Lukierski, Matscience preprint, 1964.

Theories of Particles of Arbitrary Spins

K. VENKATESAN

MATSCIENCE
Madras, India

1. INTRODUCTION

Theories of particles of spin higher than 0, $\frac{1}{2}$, and 1 have been studied ever since Dirac wrote his famous equation for the case of spin 1/2, but until recently, such studies were only of academic interest. In addition, an interest in these theories was damped for those who believed in conventional field theory because they are unrenormalizable. Considerations involving higher spins are now felt to be of practical importance by followers of the S-matrix approach with the discovery of new higher spin resonances and also perhaps in the problem of the analytic continuation in spin of the S-matrix elements and the description of the Regge poles arising therefrom.

We may approach the problem in various ways:

1. By arriving at the wave functions and the equations they satisfy by group theoretical considerations.
2. By insisting that the equation for a particle of any spin should be linear and look like the Dirac equation, and then studying the algebra of the matrices which enter it.
3. By deriving the free propagators by group-theoretic considerations and giving the Feynman rules for writing out any S-matrix element without bothering about equations of motion (or Lagrangians).
4. By the S-matrix approach where one may have to construct

invariant scattering amplitudes when a particle of higher spin is participating in the reaction.

We shall start with some elementary group-theoretical considerations. As is well known, the irreducible representations of the three-dimensional rotation group $\mathscr{D}^j$ are characterized by the quantity j which can assume only integral or half-integral values (0, $\frac{1}{2}$, 1, etc.), and represents spin. The representations for integral values of j are single-valued (tensor representation), and for half-integral values, they are double-valued (spinor representations). The rotation group is compact and is homomorphic with the group of unitary unimodular matrices in complex two-dimensional space. If there were no other considerations, we should be able to characterize particles of spin j by these representations and try to write equations for them. But the over-riding requirements of relativity, which insist that the laws of physics should be invariant under the relativistic transformations, clearly indicate that we should consider representations of the Lorentz group for our purpose, the object being to associate an equation of motion with every such representation. (This is not always in a one-to-one correspondence.)

The finite irreducible representations of the homogeneous Lorentz group $\mathscr{L}$ are obtained by its homomorphism with the group of unimodular matrices in two-dimensional complex space. That the matrices are not unitary in addition (as in the case of the rotation group) makes the representations corresponding to the dotted (complex conjugated) spinors to be distinct from those for the undotted ones. Higher dimensional spinor representations of the Lorentz group are obtained by constructing all monomials of the type

$$U_1^{2j-k} U_2^k U_{\dot{1}}^{2j'-k'} U_{\dot{2}}^{k'} \tag{1}$$

which is an element in a $(2j+1)(2j'+1)$ dimensional space and then transforming the elements by means of the two-dimensional matrices

$$M = \begin{pmatrix} a & b \\ c & d \end{pmatrix} \quad \text{and} \quad \bar{M} = \begin{pmatrix} \bar{a} & \bar{b} \\ \bar{c} & \bar{d} \end{pmatrix} \tag{2}$$

($\bar{M}$ acts only on the dotted indices.) In this way one obtains the representation $\mathscr{D}^{jj'}(M, \bar{M})$ of the Lorentz group in the $(2j+1)$ $(2j'+1)$ dimensional space. Since M and $\bar{M}$ operate only on the undotted and dotted indices, respectively, the decomposition

$$\mathscr{D}^{jj'}(M, \bar{M}) = \mathscr{D}^j(M) \otimes \mathscr{D}^{j'}(\bar{M}) \tag{3}$$

is obviously possible. (But the fact that $\bar{M}$ is not an analytic function of M prevents any possible decomposition into sums of irreducible representations so that $\mathscr{D}^{jj'}$ is really irreducible.) Just as any four-vector x has associated with it a 2×2 matrix.

$$X = \begin{pmatrix} x^0 + x^3 & x^1 - ix^2 \\ x^1 + ix^2 & x^0 - x^3 \end{pmatrix} = x^0\, 1 + \mathbf{x} \cdot \boldsymbol{\sigma} \tag{4}$$

with $\det X = x^\mu x_\mu$ every 2×2 matrix X determines a four-vector via

$$x_\mu = \tfrac{1}{2} \operatorname{Tr} (X \sigma_\mu) \tag{5}$$

Now, any unimodular 2×2 matrix, M, transforms X according to

$$X' = MXM^+ \tag{6}$$

leaving the determinant $x^\mu x_\mu$ invariant so that M indeed represents a Lorentz transformation. This enables us to write the matrix elements of X as the components of a second-rank spinor with one dotted and one undotted index $x_{\alpha\dot{\beta}} = (X)_{\alpha\dot{\beta}}$. This fact is of particular use in the construction of relativistic wave equations, since we can associate with the components of the derivative, ∂_μ, which transforms like a vector the components of spinor-derivatives of second rank:

$$\begin{aligned} \partial_{1\dot{1}} &= \partial_3 - i\partial_4 \\ \partial_{1\dot{2}} &= \partial_1 - i\partial_2 \\ \partial_{2\dot{1}} &= \partial_1 + i\partial_2 \\ \partial_{2\dot{2}} &= -\partial_3 - i\partial_4 \end{aligned} \tag{7}$$

By replacing the requirement that the laws of physics are invariant under the Lorentz transformation by the statement that these laws be invariant under the group of unimodular matrices, we arrive at the conclusion that the physical systems of interest are vectors in the space spanned by the $\mathscr{D}^{jj'}$, the spin of the particle concerned being given by $j + j'$. If, in particular, we restrict ourselves to the rotations which leave the time invariant, we obtain a direct product representation $\mathscr{D}^j \otimes \mathscr{D}^{j'}$ which except in the case of the $(j, 0)$ representation does not lead to a unique break-up thereby giving rise to $2\,s$ theories for particles of spin s.

The wave equations we shall be considering here all belong to the finite dimensional irreducible representations of the homogeneous Lorentz group. (The corresponding equations for the infinite dimensional representations some of which are unitary were studied for the first time by Dirac and Harish-Chandra.)

2. RELATIVISTIC WAVE EQUATION FOR ARBITRARY SPIN (DIRAC, FIERZ, PAULI, BELINFANTE, RARITA, AND SCHWINGER)

The type of relativistic equations we shall consider now are obtained by requiring invariance under the group of unimodular matrices, C_2. We start with some properties of the spinors that are useful. If

$$[\mathscr{E}^{rs}] = [\mathscr{E}^{\dot{r}\dot{s}}] = -[\mathscr{E}_{rs}] = -[\mathscr{E}_{\dot{r}\dot{s}}] = \begin{bmatrix} 0 & 1 \\ -1 & 0 \end{bmatrix} \tag{8}$$

then we can raise or lower the spinor indices using these, e. g.,

$$\begin{aligned} U^r &= \mathscr{E}^{rs} U_s \qquad & U_s &= \mathscr{E}_{sr} U_r \\ U^{\dot{r}} &= \mathscr{E}^{\dot{r}\dot{s}} U_{\dot{s}} \qquad & U_{\dot{s}} &= \mathscr{E}_{\dot{s}\dot{r}} U^{\dot{r}} \end{aligned} \tag{9}$$

If U and V are any two spinors, then

$$U_a V^a = \text{invariant} \qquad U_{\dot{a}} V^{\dot{a}} = \text{invariant} \tag{10}$$

$$U_a V^a = -U^a V_a \qquad \text{and} \qquad U_a U^a = -U_{\dot{a}} U^{\dot{a}} = 0 \tag{11}$$

Also

$$U^a V_a W_b + U_a V_b W^a + U_b V^a W_a = 0 \tag{12}$$

We have already mentioned that the matrix elements of the matrix X which represents a vector can also be treated as the elements of a spinor $X_{\alpha\dot{\beta}}$. The spin-tensors which effect the transition from a bispinor to a vector can be represented by

$$\begin{aligned} \sigma_{k;\dot{r}s} &\equiv [\sigma_k]_{rs} \qquad & \sigma_k^{r\dot{s}} &\equiv -[\sigma_k]_{\dot{r}s} \\ \sigma_{4,\dot{r}s} &\equiv -i[I]_{rs} \qquad & \sigma_4^{r\dot{s}} &\equiv -i[I]_{rs} \end{aligned} \tag{13}$$

The spinor indices can be raised or lowered using the $\mathscr{E}^{rs}$. Equation (7), e. g., can be rewritten as

$$\partial_{\dot{r}s} \equiv \sigma_{\mu;\dot{r}s} \partial_\mu \tag{14}$$

Also

$$\partial^{s\dot{r}} \equiv \sigma_\mu^{s\dot{r}} \partial_\mu \tag{14a}$$

In general, a tensor of degree n can be derived from a spinor of degree $2n$ as follows:

$$\psi_{\mu_1 \cdots \mu_n} = \prod_{i=1}^{n} \left(\frac{1}{2} \sigma_{\mu_i, s_i}^{\dot{r}_i} \right) \psi_{\dot{r}_1 \cdots \dot{r}_n}^{s_1 \cdots s_n} = (-1)^n \prod_{i=1}^{n} \left(\frac{1}{2} \sigma_{\mu_i}^{s_i \dot{r}_i} \right) \psi_{s_1 \cdots s_n \dot{r}_1 \cdots \dot{r}_n}, \tag{15}$$

Conversely,

$$\psi_{\dot{r}_1 \cdots \dot{r}_n}^{s_1 \cdots s_n} = \prod_{i=1}^{n} (\sigma_{\mu_i \dot{r}_i}^{s_i}) \, \psi_{\mu_1 \cdots \mu_k} \tag{16}$$

Also,

$$\begin{aligned} \partial_{\dot{u}r} \partial^{\dot{w}r} &= -\Box \delta_{\dot{u}\dot{w}} \\ \partial^{s\dot{r}} \partial_{\dot{r}t} &= -\Box \delta_{st} \end{aligned} \tag{17}$$

The simplest equation we can construct which contains a mass term which is invariant under C_2 is obviously

$$\partial_{a\dot{b}} \varphi^a = m \, \varphi_b \tag{18}$$

But this is a nonlinear equation and all free field equations are supposed to be linear. Putting $m = 0$ we get the Weyl equation

$$\partial_{a\dot{b}} \varphi^a = 0$$

or equivalently

$$(\sigma_k \partial_k + i \partial_4) \varphi = 0 \tag{19a}$$

We can circumvent the difficulty regarding nonlinearity in a different way, namely, by having a different spinor on the right-hand side of (18), i.e.,

$$\partial_{a\dot{b}} \varphi^a = im \, \chi_{\dot{b}} \tag{20}$$

But then we should also have

$$\partial^{a\dot{b}} \chi_{\dot{b}} = im \, \varphi^a \tag{21}$$

These two equations together give the Dirac equation

$$(\gamma_\mu \partial_\mu + m) \, \psi(x) = 0 \tag{22}$$

where

$$\psi(x) = \begin{vmatrix} \varphi_1(x) \\ \varphi_2(x) \\ \chi^{\dot{1}}(x) \\ \chi^{\dot{2}}(x) \end{vmatrix} \tag{23}$$

is composed of one dotted (χ) and one undotted (φ) spinors. These correspond respectively to the $\mathscr{D}^{(0, 1/2)}$ and $\mathscr{D}^{(1/2, 0)}$ representations of the Lorentz group. Thus, the Dirac spinor belongs to the representation

$$\mathscr{D}^{(1/2,0)} \dotplus \mathscr{D}^{(0,1/2)} = \mathscr{D}^{1/2} \dotplus \mathscr{D}^{1/2}$$

(for only spatial rotations)

and is a unique spin 1/2 representation, invariant under parity. (Under parity the dotted and undotted indices get interchanged.)

Dirac, Fierz, Pauli, and Belinfante extended this way of arriving at a relativistic equation to the general case of a spinor of degree n. The set of equations now are

$$\begin{aligned} \partial_{a\dot{b}} \phi^{aa_1a_2\cdots a_l}_{\dot{b}_1\dot{b}_2\cdots\dot{b}_k} &= im\chi^{a_1a_2\cdots a_l}_{\dot{b}\dot{b}_1\cdots\dot{b}_k} \\ \partial^{a\dot{b}} \chi^{a_1a_2\cdots a_l}_{\dot{b}\dot{b}_1\cdots\dot{b}_k} &= im\varphi^{aa_1\cdots a_l}_{\dot{b}_1\dot{b}_2\cdots\dot{b}_k} \end{aligned} \tag{24}$$

φ has $(l+1)$ undotted and k dotted indices and belongs to the representation $\mathscr{D}(j_1, j_1')$ with $j_1 = \frac{1}{2}(l+1)$ and $j_1' = \frac{1}{2}k$ Similarly, χ belongs to the representation $\mathscr{D}(j_2, j_2')$ with $j_2 = \frac{1}{2}l$ and $j_2^1 = \frac{1}{2}(k+1)$. The spinor built out of these

$$\psi(x) = \begin{bmatrix} \varphi(x) \\ \chi(x) \end{bmatrix}$$

belongs to the $2(2j+1)(2j'+1)$ dimensional representation

$$\mathscr{D}^{[1/2(l+1),(1/2)k]} + \mathscr{D}^{[(1/2l),1/2(k+1)]}$$

Substituting the second equation of the set [equation (24)] in the first we obtain

$$(\Box - m^2)\chi^{a_1\cdots a_k}_{\dot{b}\dot{b}_1\cdots\dot{b}_k} = 0 \tag{25}$$

Similarly,

$$(\Box - m^2)\varphi^{aa_1\cdots a_k}_{\dot{b}_1\cdots\dot{b}_k} = 0 \tag{26}$$

so that they correspond to a particle with mass m. The spin of the particle is given by

$$s = j + j' = \tfrac{1}{2}(l + k + 1)$$

Allowing l and k to run through all allowed values which satisfy this relation, we obtain $2s$ theories.

For particles of half-integer spin $s = k + 1/2$ we have the Rarita–Schwinger theory based on the $(2s+1)^2$-dimensional reducible representation

$$\left[\left(\frac{1}{2}, 0\right) + \left(0, \frac{1}{2}\right)\right] \otimes \left(\frac{2s-1}{4}, \frac{2s-1}{4}\right)$$

Introduce the quantities

$$\psi^{s}_{\mu_1\cdots\mu_r} = \prod_{i=1}^{r} \left(\tfrac{1}{2}\sigma^{\dot{u}_i}_{\mu_i;t_i}\right) \varphi^{\,st_1\cdots t_r}_{\dot{\mu}_1\cdots\dot{\mu}_r}$$

$$\psi_{\dot{k},\mu_1,\cdots,\mu_r} = \prod_{i=1}^{r} \left(\tfrac{1}{2}\sigma^{\dot{u}_1}_{\mu_i;t_i}\right) \chi^{t_1\cdots t_r}_{\dot{k},\dot{u}_1,\cdots\dot{u}_r} \tag{27}$$

which are each spinors of degree 1 and tensors of degree r and construct

$$\psi_{\mu_1\cdots\mu_r} = \begin{vmatrix} \psi^{1}_{\mu_1\cdots\mu_r} \\ \psi^{2}_{\mu_1\cdots\mu_r} \\ \psi_{\dot{1},\mu_1\cdots\mu_r} \\ \psi_{\dot{2},\mu_1\cdots\mu_r} \end{vmatrix} \tag{28}$$

This is symmetric in the suffixes $\mu_{1,\dots,}\mu_r$ and obeys the Rarita–Schwinger equation:

$$(\gamma_\mu\partial_\mu + m)\,\psi_{\mu_1\cdots\mu_r} = 0$$

Since $\phi^{st_1\cdots}_{\dot{u}_1\cdots}$ is a symmetric spinor, we have [from the second of the equations (11)],

$$\phi^{st_2\cdots t_k}_{s\dot{u}_1\cdots\dot{u}_r} = 0 \tag{30}$$

Using the relation between the γ and σ matrices, we then obtain the subsidiary condition

$$\gamma_\mu\psi_{\mu_1\mu_2\cdots\mu_r} = 0 \tag{31}$$

From equations (29) and (31) we can derive

$$(\Box - m^2)\psi_{\mu_1\cdots\mu_r} = 0$$

$$\partial_\mu\psi_{\mu\mu_2\cdots\mu_r} = 0 \qquad \psi_{\mu\mu\mu_2\cdots\mu_r} = 0 \tag{32}$$

For spin 3/2, we have

$$(\gamma_\mu\partial_\mu + m)\,\psi_\mu = 0$$

$$\gamma_\mu\psi_\mu = 0 \tag{33}$$

which yields

$$(\Box - m^2)\,\psi_\mu = 0$$

$$\partial_\mu\psi_\mu = 0 \tag{34}$$

Though the above discussion of relativistic wave equations seems to imply that they are determined exclusively by the representations of the homogeneous Lorentz group, this is not the case. In fact, Wigner has shown that the states of elementary particles are described by the unit vectors of a Hilbert space on which acts an irreducible unitary representation of the inhomogeneous Lorentz group (also called the Poincaré group), i.e., the homogeneous Lorentz group along with space and time translations. Then, the representations of this group determine all linear equations invariant under it.

The unitary irreducible representations of the Poincaré group are characterized by (a) the set of momentum vectors which can be obtained from a single momentum vector p by applying to it all proper Lorentz transformations and (b) an irreducible unitary representation of the little group which is defined as the group of all Lorentz transformations which leave any given momentum vector invariant. (a) is characterized by the Casimir operator p_μ^2 which is for time-like vectors the square of a real positive mass. The invariant which characterizes the representation of the little group is Lubanski's invariant which in the case of time-like four vectors reduces in the rest system of the paricle to m^2 times the square of the spin operator.

Wigner gives a general procedure for obtaining equations from representations. The difference of two four-vectors is invariant under displacement and transforms like a four-vector under the homogeneous Lorentz transformations. In general, we require five four-vectors to fix a frame (one to locate the origin and four to fix the orthogonal axes). One of the difference-vectors (difference of two vectors) can be identified with the momentum vector itself with the corresponding length. The other four difference vectors can be assumed to be mutually orthogonal, orthogonal to the momentum vectors and of length 1 or -1. Since this completely specifies the last difference vector in terms of the first three, we need only three independent vectors. Even the third difference vector contributes only one independent variable, since it is normalized and orthogonal to the other vectors. It turns out that this vector is also unnecessary. Hence, we are left with only two vectors, p and q. Orthogonality and normalization conditions give

$$\begin{aligned}(p \cdot p)\psi &= m^2\psi \\ (q \cdot q)\psi &= -\psi \\ (p \cdot q)\psi &= 0\end{aligned} \tag{35}$$

These are common to the equations of all representations. For an infinitesimal displacement we have

$$\psi(x + \epsilon a, p, q) = (1 - i\epsilon p \cdot a)\psi(x; p, q) \tag{36}$$

or

$$\frac{\partial \psi}{\partial x_k} = -ip^k \psi \tag{37}$$

Thus the components of x are unnecessary variables since if ψ is given as a function of p and q for the vector x, say for $x = 0$, it is determined for all other x. If we define $\psi(x, q)$ by

$$\begin{aligned} \psi(x, q) &= (2\pi)^{-2} \int d^4p \, \psi(x, p, q) \\ &= (2\pi)^{-2} \int \psi(0, p, q) \, e^{-ip\cdot x} \, d^4p \end{aligned} \tag{38}$$

it satisfies the equation

$$\frac{\partial^2}{\partial x_k \partial x^k} \hat{\psi} = m^2 \hat{\psi} \qquad q_k \frac{\partial}{\partial x_k} \hat{\psi} = 0 \tag{39}$$

For finite mass representations, the equation $(p \cdot q)\psi = 0$ restricts q to a three-dimensional space-like manifold which is perpendicular to p. Now $q_t = 0$ if $p = p^0$ which has value only along the time-axis. For an operation by an element of the little group, p^0 is unaltered and q will point toward another direction on the unit sphere, $q_x^2 + q_y^2 + q_z^2 = 1$. For the scalar representation, the representation of the little group is the identity and ψ has the same value for any orientation of q, i.e., ψ is independent of q. Thus, we have the single equation for $\psi : (p \cdot p)\psi = m^2 \psi$ which is just the Klein–Gordon equation.

We may mention, in passing, that some wave equations also possess, in addition to invariance under the Poincaré group, conformal invariance. The conformal group is a fifteen-dimensional Lie group which contains the Poincaré group and has the following physical interpretation: (1) It is the largest group of coordinate transformations of Minkowski space which leave invariant the volocity of light and (2), three of the five extra dimensions correspond to transformations to a coordinate system the origin of which is moving with uniform proper acceleration. (The two other parameters correspond to "acceleration" in the time-like direction and dilatation.) Both Maxwell's equations and the equations for the other mass-zero particles have been shown to possess conformal invariance. Segal showed how one might begin to construct a theory of elementary particles with the conformal group as the basic invariance group and how one could expect the conformal group to lead to a discrete mass spectrum and possibly a fundamental length.

3. MATRIX-ALGBERAIC APPROACH TO RELATIVISTIC WAVE EQUATIONS:

In this approach we start with a Dirac-like equation for particles of any spin, and study the algebra of the matrices appearing in them. In analogy to the Dirac equation we may write the equation for a wave function with n components ψ as

$$(\rho_\mu \partial_\mu + m\beta)\,\psi(x) = 0 \tag{40}$$

where ρ_μ and β are certain matrices of n dimensions. Substituting (46) in (47), (see below), with

$$X(\partial) = [X_{\alpha\beta}(\partial)] = \rho_\mu \partial_\mu + \mathrm{m}\beta \tag{41}$$

we get

$$m\dot{\alpha}\beta = -m^2 I \tag{42}$$

where I is the unit matrix in n dimensions. This equation shows that β is nonsingular and thus has an inverse. Multiplying (40) by β^{-1} we have (with $\beta_\mu = \beta^{-1}\rho_\mu$)

$$(\beta_\mu \partial_\mu + m)\,\psi(x) = 0 \tag{43}$$

which will be the starting point of the present approach. In order to determine the matrices β under a Lorentz transformation Λ we have

$$\psi' = \Lambda\psi \qquad \Lambda^{-1}\beta_\mu \Lambda = \alpha_{\mu\nu}\beta_\nu \tag{44}$$

For infinitesimal Lorentz transformations $\alpha_{\mu\nu} = \delta_{\mu\nu} + \epsilon_{\mu\nu}$ we have $\Lambda = (1/2)\,\varepsilon_{\mu\nu} I_{\mu\nu}$ with $I_{\mu\nu} = -I_{\nu\mu}$. Thus we have

$$[\mathrm{I}_{\rho\sigma}, \beta_\nu] = \delta_{\sigma\nu}\beta_\rho - \delta_{\rho\nu}\beta_\sigma \tag{45}$$

But these relations are not sufficient to determine completely the β-algebra. We must have the supplementary condition that all components of ψ satisfy a Klein–Gordon equation. There are two cases. Choosing the operator $X(\partial)$ as

$$X(\partial) = -(\beta_\mu \partial_\mu + m)$$

and substituting the expansion for $d(\partial)$,

$$d(\partial) = \alpha + \alpha_\mu \partial_\mu + \cdots + \alpha_{\mu_1 \cdots \mu_k} \partial_{\mu_1} \cdots \partial_{\mu_k} \tag{46}$$

into the expression

$$X(\partial)\, d(\partial) = (\square - m^2)\, I \tag{47}$$

a set of recurrence relations for the coefficients in (46) are obtained. Solving these relations we obtain the coefficients α expressed in terms of the β-matrices. Also, if l is the highest-order derivative in the expansion (46) we have

$$\alpha_{\mu_1\cdots\mu_k} = 0 \qquad \text{for } k > l \tag{48}$$

Using these facts, the following relation can be derived:

$$\Sigma^P \beta_{\mu_1}\cdots\beta_{\mu_{b-1}}[\delta_{\mu_b\mu_{b+1}} - \beta_{\mu_b\mu_{b+1}}] = 0 \tag{49}$$

where Σ^P is a summation over terms taking all possible permutations of the suffixes.

It can be proven also that the order of the differential operator, l is given by $l = 2s$ (if $m \neq 0$), where s is the maximum value of the spin of the various fields described by the field quatities (i. e., those corresponding to spins $|s|, |s-1|, \cdots$). The equation (49) then yields

$$\beta_\mu^{2s-1}(\beta_\mu^2 - 1) = 0 \tag{50}$$

(no summation over μ)

Now, if β_μ were Hermitian there would be a diagonal representation for it, the eigenvalues being $\pm 1, 0$. Such a matrix would satisfy

$$\beta_\mu(\beta_\mu^2 - 1) = 0 \tag{51}$$

Thus, equation (50) cannot be an eigenvalue equation for β_μ which cannot therefore be Hermitian if $s > 1$. Harish-Chandra showed that equation (49) does not generate a finite algebra for $s > 1$.

Next, let us consider Bhabha's theory for particles of arbitrary spin. Here it is postulated that all the information about a physical state should be obtainable from the solution of equation (43) without further subsidiary conditions unlike in the case of the Dirac–Fierz-Pauli theory. Subsidiary conditions cause trouble when interactions between particles are present. The requirement that all properties of the particles must be derivable from the field equations themselves without subsidiary conditions, makes each component of the field by itself satisfy not the Klein–Gordon equation but an equation of the type

$$\prod_{i=1}^{n}(\Box - m_i^2)\,\psi_\alpha(x) = 0 \tag{52}$$

Such a situation arises if we assume that $I_{\mu\gamma}$ does not contain any term that is a product of more than two β-matrices.

$$I_{\mu\gamma} = c\,[\beta_\mu, \beta_\gamma] \tag{53}$$

where c is a constant. Substituting this in equation (45), we have

$$\beta_\mu \beta_\lambda \beta_\gamma - \beta_\mu \beta_\gamma \beta_\lambda - \beta_\lambda \beta_\gamma \beta_\mu + \beta_\gamma \beta_\lambda \beta_\mu = \frac{1}{c}\,\delta_{\lambda\mu}\beta_\gamma - \frac{1}{c}\,\delta_{\gamma\mu}\beta_\lambda$$

$$\beta_\mu^2 \beta_\gamma + \beta_\gamma \beta_\mu^2 - 2\beta_\mu \beta_\gamma \beta_\mu = \frac{1}{c}\,\beta_\gamma \qquad \text{for } \mu \neq \gamma \tag{54}$$

(no summation over μ)

But since $-i\,I_{kl}\,(k, l = 1, 2, 3)$ behave like angular momentum operators, the equations (45) written out show that $\sqrt{c}\,\beta_\mu$ and $-iI_{kl}$ also have eigenvalues $(s, s-1, \cdots, -s)$ as $-iI_{kl}$ has. Hence, the characteristic equation for β_μ is

$$(\sqrt{c}\;\beta_\mu - s)(\sqrt{c}\;\beta_\mu - s + 1)\cdots(\sqrt{c}\;\beta_\mu + s) = 0 \tag{55}$$

or

$$[(\sqrt{c}\;\beta_4\partial_4)^2 - s^2\partial_4^2)]\,[(\sqrt{c}\;\beta_4\partial_4)^2 - (s-1)^2\partial_4^2] \times \cdots$$
$$[(\sqrt{c}\;\beta_4\partial_4)^2 - \partial_4^2]\,(\sqrt{c}\;\beta_4\partial_4)\,\psi = 0 \tag{56}$$

for an integer s. In the rest system of the particle

$$\Box = \partial_4^2 \qquad \beta_\mu\partial_\mu = \beta_4\partial_4 \tag{57}$$

Hence,

$$[s^2\Box - (\sqrt{c}\;\beta_\mu\partial_\mu)^2]\cdots[\Box - (\sqrt{c}\;\beta_\mu\partial_\mu)^2]\,(\sqrt{c}\;\beta_\mu\partial_\mu)\,\psi = 0 \tag{58}$$

which must hold in any Lorentz frame. Putting $\beta_\mu\partial_\mu = m$ and using a similar procedure for half-integral s, we have

$$\left[\left(\Box - \frac{cm^2}{s^2}\right)\left(\Box - \frac{cm^2}{(s-1)^2}\right)\cdots(\Box - (m^2)\right]\psi = 0$$

for integer s (59)

$$\left[\left(\Box - \frac{cm^2}{s^2}\right)\left(\Box - \frac{cm^2}{(s-1)^2}\right)\cdots\left(\Box - \frac{cm^2}{(1/2)^2}\right)\right]\psi = 0$$

for half-integer s

These equations show that ψ for spin $s > 1$ satisfies the multi-mass Klein–Gordon equation (52). The rest mass of particles with higher spin $(s > 1)$ can take various values, i. e., $\sqrt{c}\,m/s, \cdots, \sqrt{c}\,m$ for an integer spin, and $\sqrt{c}\;m/s\,, \cdots, 2\sqrt{c}\,m$ for half-integral s. In the case of spins 0, 1/2, and 1, this theory is the same as the previous ones.

Introducing a fifth index in the commutation relations for the infinitesimal generators for the homogeneous Lorentz group

$$[I_{\alpha\beta}, I_{\gamma\partial}] = -\delta_{\alpha\gamma} I_{\beta\partial} + \delta_{\alpha\partial} I_{\beta\gamma} + \delta_{\beta\gamma} I_{\alpha\partial} - \delta_{\beta\partial} I_{\alpha\gamma} \tag{60}$$

by writing

$$\beta_\mu = I^{\mu 5} = -I^{5\mu} \qquad g^{55} = -1 \tag{61}$$

Bhabha noticed that the commutation relations given by equations (60), (45), and (53) (with c absorbed in the β_μ's) are just the integrability conditions for the generators for infinitesimal transformations of the Lorentz group in five-dimensions, i. e., the de Sitter group.

The fact that particles of spin higher than 1 have to exist in more than one mass state makes this theory only of academic interest.

We observe in both the two cases considered, i. e., depending on whether ψ obeys a Klein–Gordon equation or equation (52), that we have for the spin 1/2 case.

$$\beta_\mu \beta_\gamma + \beta_\gamma \beta_\mu = 2\delta_{\mu\gamma} \tag{62}$$

which is the Dirac theory, and for spins 1 and 0

$$\beta_\mu \beta_\gamma \beta_\lambda + \beta_\lambda \beta_\gamma \beta_\mu = \beta_\mu \delta_{\gamma\lambda} + \beta_\lambda \delta_{\mu\gamma} \tag{63}$$

in conformity with the theory of Duffin, Kemmer, and Petiau.

4. FEYNMAN RULES FOR ANY SPIN

We shall describe now Weinberg's procedure regarding the rules for obtaining the S-matrix element of any process involving particles of arbitrary spin, considering separately the case of particles with and without mass. The procedure bypasses the introduction of a Lagrangian or an equation of motion, by making some assumptions of a general nature. First let us consider the case of particles with mass. Three assumptions are made:

(1) Use of the Dyson expansion for the S-matrix (i. e., perturbation theory)

$$S = \sum_{n=0}^{\infty} \frac{(-i)^n}{n!} \int_{-\infty}^{\infty} dx_1 dx_2 \cdots dx_n$$

$$T\{H^1(x_1) \cdots H^1(x_n)\} \tag{64}$$

where the H^1 are Hamiltonian densities in the interaction representation.

(2) Lorentz invariance of the S-matrix which is imposed by requiring that (*a*) $H^1(x)$ be a scalar, i.e., under the unitary operator $U[\Lambda, a]$ of the Poincaré group

$$U[\Lambda, a]\, H^1(x)\, U^{-1}[\Lambda, a] = H^1(\Lambda x + a) \tag{65}$$

(*b*) for $x - y$ space-like

$$[H^1(x), H^1(y)] = 0 \tag{66}$$

(This is to ensure that the θ-functions implicit in the time-ordered product are scalars.)

(3) Particle interpretation: The $H^1(x)$ are to be constructed out of creation and annihilation operators for the free particles described by H^0. This is done by building the $H^1(x)$ out of combinations of field operators which are themselves linear combinations of the creation and annihilation operators, transforming according to

$$U[\Lambda, a]\, \psi_n(x)\, U^{-1}[\Lambda, a] = \Sigma \mathscr{D}_{nm}[\Lambda^{-1}]\, \psi_m(\Lambda x + a) \tag{67}$$

(where $\mathscr{D}_{nm}(U)$ is a representation of Λ). Also,

$$[\psi_m(x), \psi_n(y)]_{\pm} = 0 \qquad (x - y) \text{ space-like} \tag{68}$$

The quantum field is used as a mere artifice in the construction of the invariant S-matrix. The procedure is to construct a physical state with a given momentum and spin, obtain the transformation properties of the creation and annihilation operators from this, construct the field operator as a linear combination of these, and finally obtain an expression for the propagator by evaluating the expectation value of the time-ordered product of such operators.

Call the particular Lorentz transformation $L(p)$ the boost which takes a particle of mass m from rest to momentum p

$$\begin{gathered} L^i_j(p) = \delta_{ij} + \hat{p}_i \cdot \hat{p}_j\, [\cosh(\theta - 1)] \\ L^i_0(p) = L^0_i(p) = \hat{p}_i \sinh\theta \qquad L^0_0(p) = \cosh\theta \\ \hat{p} = \frac{p}{|p|} \qquad \sinh\theta = \frac{|p|}{m} \\ \cosh\theta = \frac{\omega}{m} = \frac{[p^2 + m^2]^{1/2}}{m} \end{gathered} \tag{69}$$

If $|\sigma>$ is the state of a particle of mass m, spin j and z components of spin σ ($\sigma = j, \cdots, -j$) and at rest, the corresponding state of momentum p is

$$|p; \sigma> = \left[\frac{m}{\omega(p)}\right]^{1/2} U[L(p)]\, |\sigma> \tag{70}$$

where U is the unitary operator corresponding to $L(p)$ which acts in the Hilbert space of physical states. For an arbitrary Lorentz transformation Λ^{μ}_{ν}, we have

$$\begin{aligned}
U[\Lambda]\,|p;\sigma> &= \left[\frac{m}{\omega(p)}\right]^{1/2} U[\Lambda]\,U[L(p)]\,|\sigma> \\
&= \left[\frac{m}{\omega(p)}\right]^{1/2} U[L(\Lambda p)]\,U[L^{-1}(\Lambda p)\,\Lambda\,L(p)]\,|\sigma> \\
&= \left[\frac{m}{\omega(p)}\right]^{1/2} \sum_{\sigma'} U[L(\Lambda p)]\,|\sigma'> \\
&\qquad \times <\sigma'|U[L^{-1}(\Lambda p)\,\Lambda\,L(p)]|\sigma> \\
&= \left[\frac{\omega(\Lambda p)}{\omega(p)}\right]^{1/2} \sum_{\sigma'} |\Lambda\,p,\sigma'> \mathscr{D}^{j}_{\sigma'\sigma}\,[L^{-1}(\Lambda p)\cdot\Lambda L(p)]
\end{aligned} \tag{71}$$

with

$$\mathscr{D}^{j}_{\sigma',\sigma}[R] = <\sigma'|U[R]|\sigma> \tag{72}$$

R is the rotation $L^{-1}(\Lambda p)\,\Lambda L(p)$ and the $\mathscr{D}^{j}$ are the representations of the rotation group. States containing many particles can be constructed in the same way. For each particle, these states can be constructed by acting on the bare vacuum with creation operators $a^{*}(p,\sigma)$ satisfying the Bose or Fermi commutation relations.

$$[a(p,\sigma), a^{*}(p',\sigma^{1})]_{\pm} = \delta_{\sigma\sigma^{1}}\delta^{3}(p-p') \tag{73}$$

Thus, we can replace (71) by

$$U[\Lambda]\,a^{*}(p,\sigma)\,U^{-1}[\Lambda] = \left[\frac{\omega(\Lambda p)}{\omega(p)}\right]^{1/2} \sum_{\sigma'} \mathscr{D}^{j}_{\sigma\sigma'}[L^{-1}(\Lambda p)\,\Lambda L(p)] \times a^{*}(\Lambda p,\sigma') \tag{74}$$

The transformation law for the annihilation operator $a(p,\sigma)$ is [on using the unitarity of $\mathscr{D}^{j}$ and taking the adjoint of (74)].

$$U[\Lambda]\,a(p,\sigma)\,U^{-1}[\Lambda] = \left[\frac{\omega(\Lambda p)}{\omega(p)}\right]^{1/2} \sum_{\sigma'} \mathscr{D}^{j}_{\sigma\sigma'}[L^{-1}(p)\Lambda^{-1}L(\Lambda p)]a(\Lambda p,\sigma') \tag{75}$$

We can write (74) also in the form

$$\begin{aligned}
U[\Lambda]\,a^{*}(p,\sigma)\,U^{-1}[\Lambda] &= \left[\frac{\omega(\Lambda p)}{\omega(p)}\right]^{1/2} \\
&\times \sum_{\sigma'} \left\{C\mathscr{D}^{j}[L^{-1}(p)\,\Lambda^{-1}L(\Lambda p)]C^{-1}\right\} a^{*}(\Lambda p,\sigma')
\end{aligned} \tag{76}$$

where C is the $(2j+1)(2j+1)$ matrix with $C^*C=(-1)^{2j}$; $C^+C=1$. Also,

$$\mathscr{D}^j_{\sigma',\sigma}[R] = \left\{C\mathscr{D}[R^{-1}]C^{-1}\right\}_{\sigma\sigma'} \tag{77}$$

The antiparticle creation operator $b^*(p,\sigma)$ transforms as $a^*(p,\sigma)$.

So far, we have used only the matrices of the rotation group. The complexities of higher spin enter only when we try to construct a field operator which transforms simply under the homogenous Lorentz group. We cannot write the field operator direclty as a sum of the Fourier transforms of creation and annihilation operators since the latter behave under Lorentz transformation in a way dependent on the individual momentum p so that the ordinary Fourier transform would not have a covariant character. To construct fields with simple transformation properties, it is necessary to extend $\mathscr{D}^j[R]$ to a representation of the homogeneous Lorentz group so that the p-dependent factors in (75) and (76) can be grouped with the operators. There are as many ways of doing this as there are representations of the Lorentz group with the sum of the number of dotted and undotted indices adding up to $2j$.

Since

$$\mathscr{D}^j[L^{-1}(p)\Lambda^{-1}L(\Lambda p)] = [\mathscr{D}^j]^{-1}[L(p)]\,\mathscr{D}^j[\Lambda^{-1}]\,\mathscr{D}^j[L(\Lambda p)] \tag{78}$$

we can write

$$\begin{aligned} U[\Lambda]\,\alpha(p,\sigma)\,U^{-1}[\Lambda] &= \sum_{\sigma'} \mathscr{D}^j_{\sigma.\sigma'}[\Lambda^{-1}]\,\alpha(\Lambda p,\sigma') \\ U[\Lambda]\,\beta(p,\sigma)\,U^{-1}[\Lambda] &= \sum_{\sigma'} \mathscr{D}^j_{\sigma.\sigma'}[\Lambda^{-1}]\,\beta(\Lambda p,\sigma') \end{aligned} \tag{79}$$

with

$$\begin{aligned} \alpha(p,\sigma) &\equiv [2\omega(p)]^{1/2} \sum_{\sigma^1} \mathscr{D}^j_{\sigma\sigma'}[L(p)]\,a(p,\sigma^1) \\ \beta(p,\sigma) &\equiv [2\omega(p)]^{1/2} \sum_{\sigma^1} \left\{\mathscr{D}^j[L(p)]C^{-1}\right\}_{\sigma\sigma^1} b^*(p,\sigma^1) \end{aligned} \tag{80}$$

we have

$$\phi_\sigma(x) = (2\pi)^{-3/2}\int \frac{d^3p}{2\,\omega(p)}\,[\xi\alpha(p,\sigma)\,e^{ipx} + \eta\beta(p,\sigma)\,e^{-ipx}] \tag{81}$$

ξ and η are constants. This is the simplest field operator with the properties we require of it. The commutator is

$$[\phi_\sigma(x), \phi^+_{\sigma'}(y)]_\pm = \frac{m^{-2j}}{(2\pi)^3} \int \frac{d^3p}{2\omega(p)} \prod{}^{j}_{\sigma\sigma'}[p, \omega(p)] \times \left\{|\xi|^2 \exp[ip(x-y)] \pm |\eta|^2 \exp[-ip(x-y)]\right\} \tag{82}$$

where the matrix $\prod [p, \omega(p)]$ is given by

$$\begin{aligned} m^{-2j} \prod [p, \omega(p)] &= \mathscr{D}^j[L(p)]\, \mathscr{D}^j [L(p)]^+ \\ &= \exp(-2\, p \cdot J\theta) \\ &= (-)^{2j} t^{\mu_1\mu_2\cdots\mu_{2j}}_{\sigma\sigma'} p_{\mu_1}\dots p_{\mu_{2j}} \end{aligned} \tag{83}$$

where t is a constant symmetric traceless tensor and J is the infinitesimal operator of rotation. The second step in (83) arises if we choose the $\mathscr{D}(j, o)$ representation of the homogeneous Lorentz group. The momentum factors in (83) are equivalent to derivatives outside the integral in (82) and it is known that such an integral will vanish outside the light-cone if and only if the coefficients of $\exp[ip(x-y)]$ and $\exp[-ip(x-y)]$ are equal and opposite, i.e., $|\xi|^2 = \mp(-)^{2j}|\eta|^2$. There is sense only if $\mp(-)^{2j} = 1$, which gives integer spins for bosons and half-integer spins for fermions. We can choose the phase and normalization of $\phi_\sigma(x)$ such that $\xi = \eta = 1$. Thus,

$$[\phi_\sigma(x), \phi^+_{\sigma'}(y)]_\pm = i(-im)^{-2j}\, t^{\mu_1\cdots\mu_{2j}}_{\sigma\sigma'} \times \partial_{\mu_1}\partial_{\mu_2} \cdots \partial_{\mu_{2j}} \Delta(x-y) \tag{84}$$

where Δ is the usual commutator function for a massive spinless particle.

We are now in a position to give the Feynman rules. Let us choose arbitrarily the interaction density as

$$H^1(x) = g \sum_{\sigma_1\sigma_2\sigma_3} \begin{pmatrix} j_1 j_2 j_3 \\ \sigma_1\sigma_2\sigma_3 \end{pmatrix} \phi^{(1)}_{\sigma_1}(x)\, \phi^{(2)}_{\sigma_2}(x)\, \phi^{(3)}_{\sigma_3}(x) + \text{h. c.} \tag{84a}$$

Wick's rule now gives: (a) for each vertex a factor $-ig \begin{pmatrix} j_1 j_2 j_3 \\ \sigma_1\sigma_2\sigma_3 \end{pmatrix}$ (b) for each internal line running from x to y, a propagator

$$\begin{aligned} < T\left\{\phi_\sigma(x)\, \phi^+_{\sigma'}(y)\right\} >_0 &= \theta(x-y) <\phi_\sigma(x)\, \phi^+_{\sigma'}(y)>_0 \\ &\quad + (-)^{2j}\theta(y-x) >\phi^+_{\sigma'}(y)\, \phi_\sigma(x)>_0 \end{aligned} \tag{85}$$

(c) for an external line for a particle of spin j, z component μ, a wave function

$$\frac{1}{[2\,\omega(p)]^{1/2}(2\pi)^{3/2}} \mathscr{D}^j_{\sigma\mu}[L(p)] \exp(ip \cdot x)$$

for a particle destroyed

$$\frac{1}{[2\,\omega(p)]^{1/2}(2\pi)^{3/2}}\mathscr{D}^{j*}_{\sigma\mu}[L(p)]\exp(-ip\cdot x)$$

for a particle created

$$\frac{1}{[2\,\omega(p)]^{1/2}(2\pi)^{3/2}}\left\{\mathscr{D}^{j}[L(p)]C^{-1}\right\}_{\sigma\mu}\exp(-ip\cdot x)$$

for anti-particle created

$$\frac{1}{[2\,\omega(p)]^{1/2}(2\pi)^{3/2}}\left\{\mathscr{D}^{j}[L(p)]C^{-1}\right\}^{*}_{\sigma\mu}\exp(ip\cdot x)$$

for anti-particle destroyed

Here

$$\mathscr{D}^{j}[L(p)] = m^{-2j}\Pi^{j}[p^{1}] \tag{87}$$

where

$$p^{1} = \left\{\hat{p}\,[(1/2)m(\omega - m)]^{1/2},\, [(1/2)m(\omega + m)]^{1/2}\right\} \tag{88}$$

(d) integrate over all points x, y and sum over the indices σ, σ^{1}, etc.
(e) a (—) sign is to be included for each fermion loop.

Regarding the propagator there is a difficulty, namely, according to (85) above, the θ-function hangs outside the derivatives arising in the vacuum expectation value of the field operators and cannot be commuted past them as in the case of $j = 0$ or $j = 1/2$, since then the derivative acts on the θ-function as well as on the Δ^{+} and Δ^{-} functions. On the other hand, only the expression.

$$\mathrm{S}_{\sigma\sigma^{1}}(x-y) = -i(-im)^{-2j}\,\mathrm{t}^{\mu_1\mu_2\cdots\mu_{2j}}_{\sigma\sigma^{1}} \times \partial_{\mu_1}\partial_{\mu_2}\cdots\partial_{\mu_{2j}}\Delta^{c}(x-y) \tag{89}$$

has the correct behavior

$$\mathscr{D}^{j}[\Lambda]\,\mathrm{S}(x)\,\mathscr{D}^{j}[\Lambda]^{+} = \mathrm{S}(\Lambda x) \tag{90}$$

to guarantee a Lorentz-invariant S-matrix.

To remedy the situation, noncovariant contact terms have to be added to cancel out the noncovariant terms in the propagator. In a Lagrangian formalism such contact terms would be generated automatically in the transition from the interaction Lagrangian to the Hamiltonian. For our purpose, we shall merely take (89) as the correct propagator. In momentum space it reads

$$\begin{aligned}\mathrm{S}_{\sigma\sigma^{1}}(q) &= \int d^{4}x\, e^{-iq\cdot x}\,\mathrm{S}_{\sigma\sigma^{1}}(x)\\ &= (-i)(-im)^{-2j}\frac{\Pi_{\sigma\sigma^{1}}(q)}{q^{2}+m^{2}-i\epsilon}\end{aligned} \tag{91}$$

For a parity-conserving interaction we must have both the (j, o) field $\phi_\sigma(x)$ and the (o, j) field $\chi_\sigma(x)$ which can be combined into the $2(2j+1)$ component field

$$\phi_\sigma(x) = \begin{bmatrix}\phi_\sigma(x)\\ \chi_\sigma(x)\end{bmatrix} \tag{92}$$

ϕ satisfies not only the Klein-Gordon equation but also the equation

$$[\gamma^{\mu_1\mu_2\cdots\mu_{2j}}\,\partial_{\mu_1}\ldots\partial_{\mu_{2j}} + m^{2j}]\phi(x) = 0 \tag{93}$$

since

$$\begin{aligned}\overline{\Pi}(-i\partial)\,\phi(x) &= m^{2j}\chi(x)\\ \Pi(-i\partial)\,\chi(x) &= m^{2j}\phi(x)\end{aligned} \tag{94}$$

In (93)

$$\gamma^{\mu_1\cdots\mu_{2j}} = (-)2j\begin{bmatrix}0 & t^{\mu_1\cdots\mu_{2j}}\\ \bar{t}^{\mu_1\cdots\mu_{2j}} & 0\end{bmatrix} \tag{95}$$

Instead of using the (j, o) representation, we can use any representation $\mathscr{D}_{nm}[\Lambda]$ of the Lorentz group and also carry out the above construction for this case.

The Feynman rules for massless particles in the $(2j+1)$ component or $2(2j+1)$ component formalisms are the same as for $m > 0$ when the limit $m \to 0$ is taken in the wave functions and the propagators. But not all the field types which can be constructed out of the creation and annihilation operators for $m > 0$ can be so constructed for $m = 0$. Specifically, the annihilation operator for a massless particle of helicity λ and the creation operator for antiparticle with helicity $-\lambda$ can be used to form a field transforming as

$$U[\Lambda]\,\phi_n(x)\,U[\Lambda]^{-1} = \sum_n \mathscr{D}_{nm}[\Lambda^{-1}]\,\phi_m(\Lambda x) \tag{96}$$

only under those representations of the homogeneous Lorentz group for which $\lambda = B - A$. This arises from the non-semi-simple nature of the little group for $m = 0$. In this case there is no rest system. The momentum vector left invariant under the little group R_μ is $k^\nu = (0, 0, k, k)$. If $|\lambda>$ represents the set of states which furnishes a representation of the little group, the unitary operator $U[R]$ does not change the momentum of the states $|\lambda>$. For infinitesimal transformations we have

$$R^\mu_\gamma = \delta^\mu_\gamma + \Omega^\mu_\gamma \tag{97}$$

so that

$$\Omega^\mu_\gamma k^\gamma = 0 \tag{98}$$

From the property of Lorentz transformations

$$\Omega^{\mu\gamma} = -\Omega^{\gamma\mu} \tag{99}$$

The general $\Omega^{\mu\gamma}$ satisfying these two conditions is a function of the parameters θ, χ_1, χ_2 with nonzero components given by

$$\Omega^{12} = \theta \qquad \Omega^{10} = \Omega^{13} = \chi_1 \qquad \Omega^{20} = \Omega^{23} = \chi_2 \tag{100}$$

We could write the generators of the Lorentz group as

$$J = \tfrac{1}{2}\,\epsilon_{ijk}\,I_{kj} \qquad K_i = I_{i4} = -I_{4i} \tag{101}$$

The unitary operator corresponding to the little group,

$\mathscr{D}\,[R\,(\theta, \chi_1, \chi_2)]$ is obtained from that for the Lorentz group.

On using (100), we have

$$\mathscr{D}\,[R\,(\theta, \chi_1, \chi_2)] = 1 + i\theta J_3 + i\chi_1 L_1 + i\chi_2 L_2 \tag{102}$$

If we form the combinations

$$A = \tfrac{1}{2}\,[J + iK] \qquad B = \tfrac{1}{2}\,[J - iK] \tag{103}$$

then both A and B have angular-momentum commutation relations. In terms of their components (102) becomes

$$\mathscr{D}\,[R\,(\theta, \chi_1, \chi_2)] = 1 + i\theta\,(A_3 + B_3) + (\chi_1 + i\chi_2)(A_1 - iA_2) + (\chi_1 - i\chi_2)(B_1 + iB_2) \tag{104}$$

Now it is well known that if we require the states $|\lambda >$ to form a finite set, L_1 and L_2 which behave like translation operators in two-dimensional space should be such that

$$L_1|\lambda> = 0 \qquad L_2\,|\lambda> = 0 \tag{105}$$

It can be shown also by constructing the creation and annihilation parts of the field operator ϕ as linear combinations of the creation operators $a^+(p, \lambda)$ and the annihilation operators, $a(p, \lambda)$ with fixed helicity λ that if $U_m(\lambda)$ is the wave function corresponding to a prescribed momentum k,

$$\Sigma\,\mathscr{D}_{nm}[R]\;U_m(\lambda) = \exp\left\{i\lambda\theta\,[R]\right\}U_n(\lambda) \tag{106}$$

In view of (102) and (105), this yields the set of three equations

$$[A_3 + B_3]\;U\,(\lambda) = \lambda U\,(\lambda)$$

$$[A_1 - iA_2]\;U\,(\lambda) = 0$$

$$[B_1 + iB_2]\;U\,(\lambda) = 0$$

The last two equations of (107) make $U(\lambda)$ to be an eigenvector of A_3 and B_3 with eigenvalues $-A$ and B. Then, the first equation gives

$$-A + B = \lambda \tag{108}$$

For a left-handed particle with $\lambda = -j$, the various possible fields are $(j, 0), (j + 1/2, 1/2), (j + 1, 1) \cdots$. For a right-handed particle with $\lambda = j$, the representations allowed are $(0, j)$, $(1/2, j + 1/2)$, $(1, j + 1) \cdots$. If parity is conserved, then the particle must exist in both states $\lambda = \pm j$ and the field must then transform reducibly, e. g., like $(j, 0) + (0, j)$. In particular, we note that the vector representation, $(1/2, 1/2)$ corresponding to a description in terms of the scalar and vector potentials is group-theoretically the wrong representation for the photon field while the $(1, 0)$ or $(0, 1)$ representation which is a correct representation corresponds to a description in terms of field strengths. But for computation purposes it is the first mode of description which is convenient.

5. ARBITRARY SPINS AND THE S-MATRIX APPROACH

The statement of relativistic invariance of scattering amplitudes is that the amplitude remains invariant when the momentum and spin variables of each particle are transformed according to the corresponding irreducible unitary representation of the inhomogeneous Lorentz group. To construct such an amplitude is to find the most general function that has the required transformation properties. For outgoing particles and incoming antiparticles with spin and four-momenta s_i, k_i and incoming particles and outgoing antiparticles with spin and four-momenta s_j, k_j all with nonzero rest masses (the case for zero rest mass which involves the use of "null spinors" has been considered by Zwanziger), the S-matrix transforms as

$$S(k) = \bigotimes_i \mathscr{D}^{s_i}[A'_i(-k_i)] \otimes \bigotimes_j \mathscr{D}^{s_j}[A'_j(k_j)]^* \, S[\Lambda^{-1}(A)k] \tag{109}$$

where

$$A'(k) = B^{-1}_{k\to p} A B_{q\leftarrow p} \qquad \Lambda(A^{-1})k = q \tag{110}$$

k stands for the set of incoming and outgoing four momenta with $\sum_n k_n = 0$.

$$\Lambda\,(B_{k\leftarrow p})\,p = k \tag{111}$$

The matrix A' transforms p into itself where $p = (m, 0, 0, 0)$ and, hence, is a rotation

$$B_{k\leftarrow p}\, B^{+}_{k\leftarrow p} = \frac{k^{\mu}\sigma_{\mu}}{m} \tag{112}$$

the solution of which is $B_{k\leftarrow p} = A_{k\leftarrow p}$, where $A_{k\leftarrow p}$, is the hermitian matrix $\left(\frac{k, \sigma}{m}\right)^{1/2}$

Using the fact that

$$\mathscr{D}^{s}\,[A'(k)] = \mathscr{D}^{(s,0)}\,[A'(k)] = \mathscr{D}^{(0,s)}\,[A(k)] \tag{113}$$

we have

$$\mathscr{D}^{s}\,(B^{-1}_{k\leftarrow p}\,A\,B_{q\leftarrow p}) = [\mathscr{D}^{(s,0)}\,(B_{k\leftarrow p})]^{-1} \times \mathscr{D}^{(s,0)}\,(A)\,\mathscr{D}^{(s,0)}\,(B_{q\leftarrow p}) \tag{114}$$

Introduce the Stapp M-matrix,

$$M\,(k) = \bigotimes_i \mathscr{D}^{(s_i,0)}\,(B_{k_i\leftarrow p_i}) \otimes \bigotimes_j \mathscr{D}^{(s_j,0)*}\,(B_{k_j\leftarrow p_j})\,R\,(K) \tag{115}$$

where $R = S - I$. Then M has the simple transformation law

$$M\,(K) = \bigotimes_i \mathscr{D}^{(s_i,0)}\,(A) \otimes \bigotimes_j \mathscr{D}^{(s_j,0)*}\,(A)\,M\,[\Lambda\,(A^{-1})\,K] \tag{116}$$

In general, we can write

$$M\,(K) = \sum_i A^i(K)\,Y^i\,(K) \tag{117}$$

where $A^i\,(K)$ are Lorentz scalars and the $Y^i\,(K)$ are the basis functions in spin space which have the same transformation property as the M functions, while the $A^i\,(K)$ are required to have the same singularities as the M functions. If the basis functions are polynomials in the components of the linear momenta, the scalar amplitudes can still have kinematical poles at various degenerate points where the basis functions become linearly dependent. The basis functions are built up by adding a set of spin 1/2 matrices σ_μ using the Clebsch–Gordan coefficients in a process corresponding to addition of spin. For two-body reactions, one then combines the spin matrices with tensors formed from the four-momenta to obtain a set of basis functions.

REFERENCES

[1] I. M. Gelfand, R. A. Minlos, and Z. Ya. Shapiro, "Representations of the Rotation and Lorentz Groups and Their Applications," Pergamon Press, Oxford (1963).

[2] P. Roman, *The Theory of Elementary Particles*, North-Holland Publishing Co., Amsterdam (1961).

[3] P. A. M. Dirac, *Proc. Roy. Soc.* **183**A: 284 (1945).

[4] Harish-Chandra, *Proc. Roy. Soc.* **189**A: 372 (1947).

[5] H. Umezawa, *Quantum Field Theory*, North Holland Publishing Co., Amsterdam (1956).

[5a] Y. Takahashi and H. Umezawa, *Progr. Theoret. Phys.* **9**: 14 (1953).

[5b] H. Umezawa and A. Visconti, *Nucl. Phys.* **1**: 348 (1956).

[5c] Y. Takahashi and H. Umezawa, *Nucl. Phys.* **51**: 193 (1956).

[6] E. P. Wigner, *Theoretical Physics*, International Atomic Energy Agency, Vienna (1963), p. 59.

[7] L. Gross, *J. of Math. Phys.* **5**: 687 (1964).

[8] I. E. Segal, *Duke Math. J.* **18**: 221 (1951).

[9] H. J. Bhabha, *Rev. Mod. Phys.* **17**, **21**: 200, 451 (1945), (1949).

[10] S. Weinberg, *Phys. Rev.* **133B**, **134B**: 1318, 882, (1964), (1964).

[11] A. O. Barut *et al.*, *Phys. Rev.* **130**: 442 (1963).

[12] D. Zwanziger, *Phys. Rev.* **133B**: 1036 (1964).

Bethe–Salpeter Equation and Conservation Laws in Nuclear Physics

W. BRENIG

MAX PLANCK INSTITUTE FOR PHYSICS
Munich, Germany

1, INTRODUCTION

Many of the low-energy properties of systems of strongly interacting particles can be described very well if they are considered as systems of weakly interacting quasi-particles. This description has been very fruitful in the theory of solids as well as liquid He^3 and He^4.

Recently, great progress has been made in generalizing the theory of Fermi liquids to cover the theory of finite nuclei [see A.B. Migdal, *Nucl. Phys.* 57: 29 (1964)].

Although the method of interacting quasi-particles in many cases does not lead to new results as far as applications are concerned, it sheds a new light on many older calculations or gives a theoretical justification of the quasi-particle description:

1. *The Hartree-Fock approximation.* Essentially, is this a weak-interaction approximation, assuming that $V_0/(k_f^2/2m) \ll 1$, where V_0 is a measure of the strength of the two-body interaction and $k_f^2/2m$ a measure of the average kinetic energy of the particles, k_f being the Fermi momentum.

2. *T-matrix and random-phase approximations.* Basically, these are low-density or high-density approximations assuming $ak_f \ll 1$

or $me^2/k_f \ll 1$, respectively, where a is an effective two-particle scattering length.

3. *Application of Bethe–Salpeter techniques.* In this approach one considers a sufficiently narrow energy range $\Delta\mathcal{E}$ and singles out those diagrams in a perturbation expansion which vary strongly within $\Delta\mathcal{E}$ from the ones which can be treated as constants. Usually, no attempt is made to calculate these constants. They are considered as microscopic parameters of the system. Since $\Delta\mathcal{E}/\mathcal{E}$ can be made arbitrarily small by confining oneself to a sufficiently narrow range of excitation energies, this approach is virtually exact.

However, the number of free parameters in such an approach usually is rather large. The aim of this lecture is to show that invariances such as gauge invariance, Galilean invariance, or the corresponding conservation laws following from them lead to many relations between these parameters, thus reducing the number of independent parameters considerably.

2. THE RESPONSE FUNCTION

There is a large class of experiments which can be described by means of the so-called response function. Let us consider some examples.

1. *Dipole γ-absorption.* The cross section for the absorption of a dipole γ-ray can be written as

$$\sigma(\omega) = 4\pi e^2 \omega \operatorname{Im} R(\omega) \qquad \omega \geqslant 0 \tag{1}$$

where $R(\omega)$ is the dipole response function

$$R(\omega) = \left\langle D \frac{1}{H - E_0 - \omega - i\gamma} D + D \frac{1}{H - E_0 + \omega - i\gamma} D \right\rangle \tag{2}$$

where D is the dipole operator.

2. *Elastic γ-ray scattering.* The cross section in this case is given by

$$\sigma(\omega) = \frac{8\pi}{3} \left| \frac{e^2}{m} Z - \frac{1}{2\pi^2} \int_0^\infty \sigma \, d\omega - e^2 \omega^2 R(\omega) \right|^2 \tag{3}$$

3. *Inelastic electron scattering.* As an example consider the longitudinal-dipole form factor $F_1(k)$

$$|F_1(k)|^2 = i2\pi \langle j_1(kr) \delta(H - E_0 - \omega) j_1(kr) \rangle \tag{4}$$

Further examples may be found in Migdal's paper cited above.

The quantities considered in our examples can all be obtained as special "expectation values" of a general matrix response function

$$\underset{\alpha\lambda,\mu\beta}{R(\omega)} = \left\langle \rho_{\alpha\mu} \frac{1}{H - E_0 - \omega - i\gamma} \rho^{+}_{\beta\lambda} + \rho^{+}_{\beta\lambda} \frac{1}{H - E_0 + \omega - i\gamma} \rho_{\alpha\mu} \right\rangle \tag{5}$$

where

$$\rho^{+}_{\alpha\mu} = \psi^{+}_{\alpha}\psi_{\mu} \tag{6}$$

and ψ^{+}_{α}, ψ_{α} are the creation and annihilation operators of fermions in the state characterized by the quantum number(s) α. For instance, the dipole response is obtained by

$$R(\omega) = \sum D^{*}_{\alpha\mu} R_{\alpha\lambda,\mu\beta}\, D_{\beta\lambda} \tag{7}$$

where $D_{\alpha\mu}$ are the matrix elements of the dipole operator, i.e.,

$$D = \sum D_{\alpha\mu}\rho^{+}_{\alpha\mu} \tag{8}$$

Therefore, we consider the general response function in more detail.

1. *Hartree-Fock approximation.* The shell model may serve as a first approximation for the determination of the response function. If the states defined by the quantum numbers $\alpha_1\mu$ are chosen so as to coincide with the single-particle states of the shell model, the response function takes a particularly simple form

$$R_{\alpha\lambda,\mu\beta} = R^{s0}_{\alpha\kappa}\delta_{\alpha\beta}\delta_{\mu\lambda} \tag{9}$$

where

$$R^{s0}_{\alpha\kappa}(\omega) = \frac{(1 - N_{\alpha})N_{\kappa}}{\varepsilon^{0}_{\alpha} \quad \varepsilon^{0}_{\kappa} - \omega - i\gamma} \tag{10}$$

$\varepsilon^{0}_{\alpha}, \varepsilon^{0}_{\kappa}$ are the Hartree-Fock single-particle energies of the shell model states α, κ and N_{α}, N_{κ} their occupation numbers in the groundstate

$$N_{\alpha} = \begin{cases} 1 & \text{if } \alpha \text{ is occupied} \\ 0 & \text{if } \alpha \text{ is unoccupied} \end{cases} \tag{11}$$

Equation (10) describes the fact that the particle-hole pair created by $\rho^{+}_{\alpha\kappa}$ remains in its state and has the excitation energy $\varepsilon^{0}_{\alpha} - \varepsilon^{0}_{\kappa}$.

2. *Random-phase approximation.* Usually of there are many particle-hole states of approximately the same energy as in the shell model. Because of this approximate degeneracy, the residual interaction will lead to a strong mixing of the shell-model excitations. The effect of this mixing may be treated in the so-called random-phase

approximation. In this approximation one can derive the following equation for the response function:

$$R_{\alpha\lambda,\mu\beta} = R^{s0}_{\alpha\mu}(\delta_{\alpha\beta}\delta_{\mu\lambda} - \sum V^0_{\alpha\kappa,\mu\gamma}R_{\gamma\lambda,\kappa\lambda}) \tag{12}$$

3. *Higher-order corrections.* To take into account all higher-order corrections it is convenient to use Feynman diagrams. The iterative solution of (12) may be written in matrix form as

$$R = R^{s0}(1 - V^0 R) = R^{s0} - R^{s0}V^0R^{s0} + \cdots \tag{13}$$

and illustrated by diagrams in the following manner:

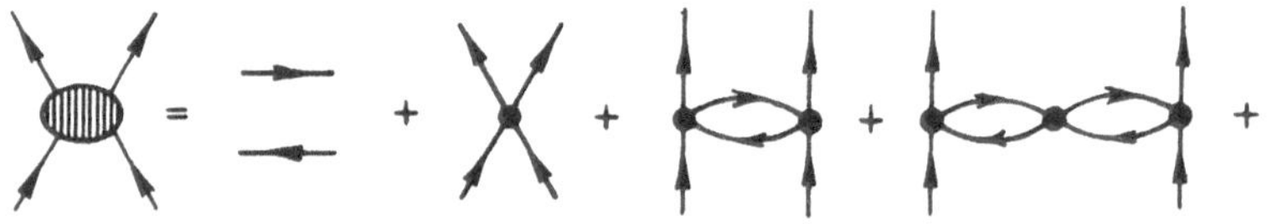

All higher-order diagrams now may be classified as either "single-particle dressing" or "vertex renormalization" diagrams. Typical single-particle dressing diagrams are

All these diagrams are summed up if one replaces the single particle propagator of the shell model G^0, indicated by an open arrow,

$$G^0_\alpha(\varepsilon) = \frac{1 - N_\alpha}{\varepsilon - \varepsilon^0_\alpha + i\gamma} + \frac{N_\alpha}{\varepsilon - \varepsilon^0_\alpha - i\gamma} \tag{14}$$

by the dressed propagator G, indicated by a full arrow,

$$G_{\alpha\beta}(\varepsilon) = Z_\alpha g_\alpha(\varepsilon)\delta_{\alpha\beta} + G^r_{\alpha\beta}(\varepsilon) \tag{15}$$

where $g(\varepsilon)$ is of the same form as G^0, ε^0_α being replaced by ε_α and G^r is nonsingular at the renormalized single-particle energies.
Typical vertex renormalization diagrams are

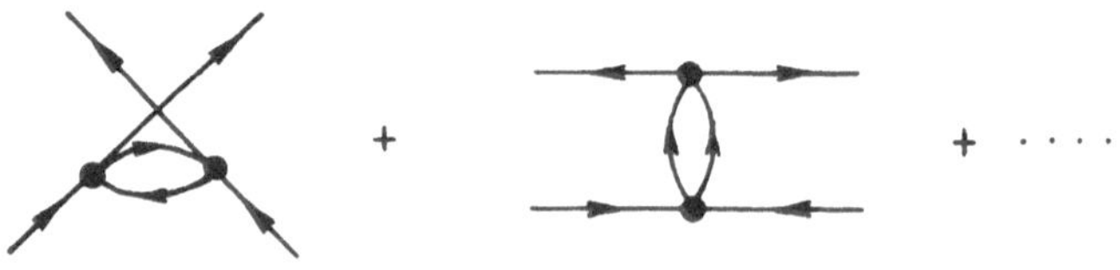

All these diagrams are summed up if one replaces the bare vertex V^0 by the renormalized vertex V. The equation for the response function for the external field

$$QRQ = QR^s(1 - VR)Q = QR^s T \tag{16}$$
$$T = Q - VR^s T$$

with

$$R^s_{\alpha\lambda,\mu\beta}(\varepsilon, \omega) = G_{\alpha\beta}\left(\varepsilon + \frac{\omega}{2}\right) G_{\lambda\mu}\left(\varepsilon - \frac{\omega}{2}\right) \tag{17}$$

and the energies of the external lines are chosen according to the diagram

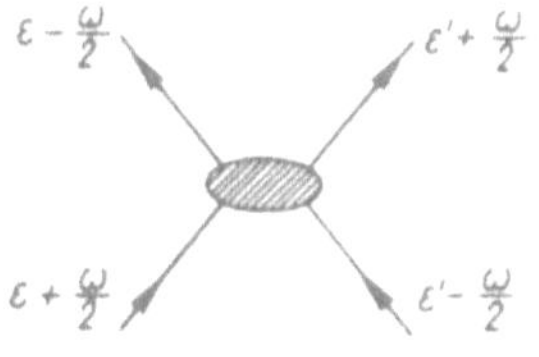

Therefore, R depends on $\varepsilon, \varepsilon'$, and ω. The $R(\omega)$ of equation (5) is obtained from $R(\omega, \varepsilon, \varepsilon')$ of equation (16) after integrating over ε and ε'.

3. TRANSFORMATION TO QUASI-PARTICLES

The crucial point in the further treatment of (16) is the fact that R^s contains a part, which is singular at the particle-hole excitation energies $\varepsilon_\alpha - \varepsilon_\kappa$. If one uses (15) one obtains

$$R^s_{\alpha\lambda,\mu\beta} = Z_\alpha Z_\beta r^s_{\alpha\mu}(\omega)\delta\left(\frac{\varepsilon_\alpha + \varepsilon_\mu}{2} - \varepsilon\right) + R^r_{\alpha\lambda,\mu\beta} \tag{18}$$

where r^S has the same form as R^{s0} but the ε^0_α's are replaced by the ε_α's. R contains no more poles and δ-functions and V similarly is nonsingular at the particle-hole excitation energies. One can now eliminate the renormalization factors z as well as R^γ. After some algebra one finds

$$\text{Im}\, QRQ = \text{Im}\, qr^s \tau \tag{19}$$

with

$$\tau = q - vr^s \tau$$

where we have introduced the quasi-particle quantities r^s, q, v, and τ instead of the renormalized ones R^s, Q, V, and J. In matrix notation they are related by

$$\begin{aligned} r^S &= Z^{-2}R^S - R^r \\ \tau &= ZT \\ q &= Z(1 + VR^r)^{-1}Q \\ v &= Z^2(1 + VR^r)^{-1}V \end{aligned} \tag{20}$$

With the aid of (19) the response function is expressed in terms of the quasi-particle quantities. The parameters contained in them are the matrix elements q of the external field, the renormalized single-particle energies $\mathcal{E}_\alpha$, and the effective vertex v. The question then arises how they can be determined.

First of all, one sees from the spectral decomposition of G that the single-particle energies of the A particle system can be read off from the energy levels of the $A + 1$ and $A - 1$ particle systems. They may therefore be taken from experiment.

We shall see now that in many cases of interest the requirement of gauge invariance leads to $Q = q$, and a combination of gauge invariance and Galilean invariance leads to further relations between the quantities q, v, and the $\mathcal{E}_\alpha$. In some cases these relations are sufficient to determine all free parameters in terms of the experimentally observable $\mathcal{E}_\alpha$'s.

4. CONSERVATION LAWS

To explore the consequences of gauge invariance, it is convenient to assume that the particles have an infinitesimal charge δ_e and then investigate the response to a purely longitudinal vector potential such as

$$Q = \delta_e(\mathbf{p}\cdot\mathbf{k} - \omega)e^{i(k\cdot r - \omega t)} \tag{21}$$

The only effect of this field is to multiply the field operators $\psi(x)$ by the phase factors exp $(i\delta\, e \cdot e^{ikx})$. After some algebra (compare, for instance, Migdal, *loc. cit.*), one can from this deduce an identity for the corresponding τ_Q in matrix notation

$$\tau_Q = e^{i\mathbf{k}\cdot\mathbf{r}} g^{-1}_{(\mathcal{E}+\omega/2)} - g^{-1}_{(\mathcal{E}-\omega/2)} e^{i\mathbf{k}\cdot\mathbf{r}} \tag{22}$$

This identity is valid for all $\mathbf{k}$ and ω. Expansion into powers of $\mathbf{k}$ after comparison of the corresponding powers on both sides of (22) yields a set of identities for the various multipoles. In particular, for the dipole term (putting $\mathbf{r} = D$),

$$(\tau_P - \omega\tau_D)_{\mu\alpha} = Z^{-1}(\mathcal{E}_\alpha - \mathcal{E}_\mu + \omega)iD_{\mu\alpha} \tag{23}$$

The right-hand side of this equation may now be inserted into equation (16) for τ. After applying the procedure which transforms (16) into the quasi-particle language (19), one finally obtains

$$(\varepsilon_\mu - \varepsilon_\alpha + \omega)D_{\alpha\mu} = \left(\frac{i}{m} P_{\alpha\mu} + \omega d_{\alpha\mu}\right) + \sum_{\beta,\lambda} v_{\alpha\lambda,\mu\beta}(N_\beta - N_\lambda)D_{\beta\lambda} \tag{24}$$

Since this is valid for all k and ω, one has the two equations

$$D_{\alpha\mu} = d_{\alpha\mu} \tag{25}$$

and

$$(\varepsilon_\mu - \varepsilon_\alpha)D_{\alpha\mu} = \frac{1}{\mathrm{m}} P_{\alpha\mu} + \sum v_{\alpha\lambda,\mu\beta}(N_\beta - N_\lambda)D_{\beta\lambda} \tag{26}$$

In an analogous way, one may derive from Galilean invariance the two corresponding equations

$$P_{\alpha\mu} = p_{\alpha\mu} \tag{27}$$

and

$$(\varepsilon_\mu - \varepsilon_\alpha)P_{\alpha\mu} = \sum v_{\alpha\lambda,\mu\beta}(N_\beta - N_\lambda)P_{\beta\lambda} \tag{28}$$

This means that in the case of n approximately degenerate levels $\varepsilon_\alpha - \varepsilon_\mu$ contributing to the response function, one has $2n$ relations among the matrix elements of the vertex operator v.

The above considerations apply to excitations with isospin zero. In the case of isospin one, Galilean invariance has no effect and the consequences of gauge invariance are modified because of the presence of exchange forces. If one uses the approximation of "effective charges," i.e., $eD = e'd$, one can relate e'/e to the exchange correction of the dipole sum rule [see W. Brenig, "Theory of Giant Dipole Resonance," *Advances in Theoretical Physics*, Vol. 1 (1965), p. 59].

On a Class of Non-Markovian Processes and Its Application to the Theory of Shot Noise and Barkhausen Noise

S. K. SRINIVASAN

INDIAN INSTITUTE OF TECHNOLOGY
Madras, India

1. INTRODUCTION

Stochastic processes arising from overlapping pulses have been the center of interest for many years in many physical problems involving noise. A naive approach to such problems consists in assuming the statistical independence of pulses generated at different times and then studying the response function which is a sum of determinate functions of the random times at which the pulses have been generated. However, in the case of noise problems, there is ample experimental evidence to indicate the crude nature of the approximation. In fact, the pulses have a fairly good correlation, particularly those whose times of generation are not separated by very large intervals of time. To be precise, we may say that the stochastic process governing the distribution of the pulse numbers is essentially non-Markovian in character, and earlier studies relating to the study of the response function are based on the simple Markovian nature of a process. Recently, we have attempted to remove the restriction and explain a certain class of noise problems on the basis of a non-Markovian model governing the pulse generation (see Refs. 1–3). The present discussion will be confined to the particular model and we shall see how a number of physical phenomena and, in particular, shot noise and Barkhausen noise can be explained in terms of this model.

2. A NON-MARKOVIAN PROCESS

Let us consider a stochastic process of the inhomogeneous Poisson type. We shall assume that the parameter $\lambda(t)$ characterizing the process is a nonnegative, continuous, and bounded function of t with respect to which the process progresses. The probability that an event happens between t and $t + dt$, is λdt, while the probability that n events occur between 0 and t, is given by

$$P(n, t) = e^{-\Lambda(t)} \frac{[\Lambda(t)]^n}{n!} \tag{1}$$

This type of Poisson process can be used to describe a number of physical phenomena such as electron emission in a counter and age dependent birth and death processes (see for example Ref. 4). In all these processes the parameter $\lambda(t)$ depends only on t and does not depend on either the number of events that have occurred before t or the points at which the events have occurred. Processes in which $\lambda(t)$ depends on n, the number of events that have occurred between 0 and t, have received some attention (see Ref. 5). However, there are quite a number of physical phenomena where $\lambda(t)$ depends not only on the number of events that have occurred prior to t but also on the positions of t. An example of this kind of process is the shot effect,[6] where an electron arriving at the anode between t_1 and $t_1 + dt_1$ diminishes the probability of any further arrival at a later time. The present discussion is confined to a particular model which has been primarily motivated by shot noise.

More specifically, let us assume that an event occurring between t and $t + dt$ diminishes the probability of the occurrence of any further event at a later time by $b \exp[-a(t' - t)]$. The process is switched on at $t = 0$ when the probability of the occurrence of an event is λ_0 (a constant). Thus, the first event happens between t_1 and $t_1 + dt_1$ with probability $e^{-\lambda_0 t_1} \lambda_0 dt_1$ while the probability of occurrence of the second event between t_2 and $t_2 + dt_2$ is given by

$$P_2(t_1, t_2)\, dt_2 = \left\{\exp - \int_{t1}^{t2} \left[\lambda_0 - b\, e^{-a(t'-t_1)}\right] dt'\right\}$$
$$\left\{\lambda_0 - b\, e^{-a(t_2-t_1)}\right\} \qquad (2)^*$$

Let us denote the parameter characterizing the process by $\lambda(t)$ in analogy with the inhomogeneous Poisson process. The parameter $\lambda(t)$ *is no longer a deterministic* function of t but depends on the various random values of t at which the events have occurred. A typical realized value of $\lambda(t)$, corresponding to the events that have occurred at $t_1, t_2, \ldots, t_n$, is given by

$$\lambda^R(t) = \lambda_0 - b \sum_{i=1}^{n} \exp\left[-a(t - t_i)\right] \qquad (3)$$

The probability measure corresponding to the above realized value can be calculated using (3).

For such a process, a number of questions can be raised. However, in view of the complexity of the problem arising from the non-Markovian behavior of the process, we shall be content with the following:

1. What is the probability frequency function of λ ?
2. What is the correlation of events occurring on the t-axis ?

Let $\pi(\lambda, t)$† be the probability frequency function of $\lambda(t)$ so that $\pi(\lambda, t)\, d\lambda$ denotes the probability that has a value between λ and $\lambda + d\lambda$ at t. We proceed to obtain the Kolmogorov equation (see Ref. 5) satisfied by $\pi(\lambda, t)$. Let us increase t by Δ. Between t and $t + \Delta$ either an event occurs or it does not. In the former case, $\lambda(t)$ increases deterministically during the interval $(t, t + \Delta)$ as can be seen from (2), the rate of increase being given by

$$\frac{d\lambda}{dt} = a\,(\lambda_0 - \lambda) \qquad (4)$$

If, on the other hand, an event occurs between t and $t + \Delta, \lambda(t)$, suddenly diminishes by b. Using these results, we obtain

$$\pi(\lambda, t + \Delta) = (1 - \lambda\Delta)\,\pi\,(\lambda - a\,\overline{\lambda_0 - \lambda}\,\Delta, t)\, d\,(\lambda - \overline{\lambda_0 - \lambda}\, a\Delta)$$
$$+ \pi(\lambda + b, t)\,(\lambda + b)\,\Delta\, d(\lambda + b) + O(\Delta) \qquad (5)$$

---

* $P_2(t_1, t_2)$ is a conditional probability frequency function.

†Throughout, we shall use the symbol π to denote any probability frequency function, the distinction between two different probability frequency functions being apparent from the context.

Proceeding to the limit as $\Delta \to 0$, we obtain

$$\frac{\partial \pi(\lambda, t)}{\partial t} = (a - \lambda)\,\pi(\lambda,t) - a(\lambda_0 - \lambda)\frac{\partial \pi(\lambda,t)}{\partial \lambda} + (\lambda + b)\,\pi\,(\lambda + b,t) \tag{6}$$

Equation (6) is true only if $\lambda > 0$. When $\lambda < 0$, it is easy to see that $\pi(\lambda, t)$ satisfies the equation

$$\frac{\partial \pi(\lambda,t)}{\partial t} = a\pi(\lambda,t) - a(\lambda_0 - \lambda)\frac{\partial \pi(\lambda,t)}{\partial \lambda} + (\lambda + b)\,\pi\,(\lambda + b,t) \tag{7}$$

However, in such a case λ cannot be a probability magnitude. This difficulty can be overcome by defining λ' by

$$\begin{aligned} \lambda' &= \lambda && \text{for } \lambda > 0 \\ &= 0 && \text{otherwise} \end{aligned} \tag{8}$$

We observe that it is λ' that has probability significance, and in any problem we have to deal with only the moments of λ'. Equation (6) is very difficult to solve explicitly for $\pi(\lambda,t)$. However, it is possible to obtain the moments of λ'. Defining

$$p\,(n, t) = \int_{-b}^{\infty} \pi\,(\lambda', t)\,\lambda'^n\,d\lambda' \tag{9}$$

We obtain

$$\frac{\partial p(n,t)}{\partial t} = -na\,p(n,t) + na\lambda_0\,p(n - 1,t) + \sum_{i=1}^{n}\binom{n}{i}p(n - i + 1, t)(-b)^i \tag{10}$$

with the conditions

$$p\,(0, t) = 1 \qquad p\,(n, 0) = \lambda_0^n \tag{11}$$

The first few moments can be explicitly calculated:

$$p\,(1, t) = \frac{a\lambda_0}{(a + b)} + \frac{b\lambda_0 \exp\,[-(a + b)\,t]}{(a + b)} \tag{12}$$

$$p(2, t) = \frac{(2\lambda_0 - a - 2b)b^2\lambda_0 \exp\,[-2(a+b)t]}{2(a + b)^2} + \frac{2b\lambda_0(2a\lambda_0 + b^2)\exp\,[-(a + b)t] + a\lambda_0(2a\lambda_0 + b^2)}{2(a + b)^2} \tag{13}$$

The moments have a simple form if we proceed to the limit as t tends to infinity. Defining $p(1)$ and $p(2)$ as the limit of $p(1, t)$ and $p(2, t)$ we find

$$p(1) = \frac{a\lambda_0}{(a+b)} \tag{14}$$

$$p(2) = [p(1)]^2 + \frac{b^2 p(1)}{2(a+b)} \tag{15}$$

In the special case $b = 0$, we find

$$p(n, t) = p(n) = \lambda_0^n \tag{16}$$

Equation (16) is consistent with the fact that the parameter λ is no longer random and that its value at any time should be equal to its initial value λ_0.

To obtain the correlation of events on the t-axis, we shall use the correlation functions that have been recently formulated (see Ref. 7) for the description of stochastic point processes.

Let $f_m(t_1, t_2, \ldots, t_m)$ be the product density of events of degree m in t-space. Then, $f_m dt_1 dt_2, \ldots, dt_m$ represents the joint probability that an event occurs between t_1 and $t_1 + dt_1$, an event between t_2 and $t_2 + dt_2$, and an event between t_n and $t_n + dt_n$ irrespective of the number of events occurring elsewhere. If we denote by $n(t)$ the number of events that have occurred between 0 and t, the mth moment of $n(t)$ is given by

$$\mathcal{E}\left\{[n(t)]^m\right\} = \sum_{i=1}^{m} C_i^m \int_0^t \int_0^t \cdots \int_0^t f_i(t_1, t_2, \ldots, t_i) dt_1 dt_2 \ldots dt_i \tag{17}$$

where C_i^m are known coefficients that have nothing to do with any particular process. Thus the moments of $n(t)$ can be obtained from the above correlations of events. In this section we shall be concerned with the explicit evaluation of the correlations of the first few orders.

We note that to obtain $f_1(t)$ a knowledge of $\pi(\lambda, t)$ is necessary. Using elementary probability arguments, we find that

$$f_1(t)dt = \int_0^\infty \pi(\lambda, t)\, d\lambda\, \lambda dt \tag{18}$$

Thus, we have

$$f_1(t) = p(1, t) \tag{19}$$

The mean number of events that have occurred between 0 and t is given by

$$\mathcal{E}\left\{n(t)\right\} = \int_0^t f_1(t)dt$$

$$= \frac{a\lambda_0 t}{a+b} + \frac{b\lambda_0}{(a+b)^2}(1 - e^{-(a+b)t}) \tag{20}$$

To obtain the mean square number of events, we must obtain $f_2(t_1, t_2)$. In view of the non-Markovian nature of the process, it is convenient to introduce the function $\pi(\lambda_2, t_2|\lambda_1, t_1)$, where $\pi(\lambda_2, t_2|\lambda_1, t_1)d\lambda_2$ represents the probability that $\lambda(t_2)$ has a value between λ_2 and $\lambda_2 + d\lambda_2$ at t_2 given that $\lambda(t)$ had a value λ_1 at t_1 and a value λ_0 initially. It is easy then to find

$$f_2(t_1, t_2) = \int_{\lambda_1}\int_{\lambda_2} \pi(\lambda_1, t_1)\, d\lambda_1\, \pi(\lambda_2, t_2|\lambda_1 - b, t_1)\, \lambda_1\lambda_2\, d\lambda_2$$

$$= \int_{\lambda_1} \mathcal{E}\left\{\lambda(t_2)|\lambda_1 - b, t_1\right\} \lambda_1\, \pi(\lambda_1, t_1)\, d\lambda_1 \tag{21}$$

where $\mathcal{E}\{\lambda(t_2)|\lambda_1 - b, b_1\}$ represents the conditional moment of λ at t_2 given that λ had a value $\lambda_1 - b$ at t_1. We can evaluate $\mathcal{E}\{\lambda(t_2)|\lambda_1 - b, t_1\}$ if we know the conditional probability frequency function $\pi(\lambda_2, t_2|\lambda_1, t_1)$ where $\pi(\lambda_2, t_2|\lambda_1, t_1)\, d\lambda_2$ denotes the probability that $\lambda(t_2)$ has a value between λ_2 and $\lambda_2 + d\lambda_2$ given that $\lambda(t_1) = \lambda_1$. Then, π is determined by setting up the Kolmogorov differential equations, and, as the calculations are long, we give only the final result. (Details can be found in Refs. 1 and 2.)

$$f_2(t_1, t_2) = p(2, t_1)e^{-(a+b)(t_2-t_1)}$$
$$+ p(1, t_1)\left[\frac{a\lambda_0(1 - e^{-(a+b)(t_2-t_1)}}{(a + b)} - b\, e^{-(a+b)(t_2-t_1)}\right] \tag{22}$$

An interesting feature that emerges from (22) is the existence of the limit of $f_2(t_1, t_2)$ where both t_1 and t_2 tend to infinity in such a manner that $t_2 - t_1$ remains a constant τ. Thus we have

$$\lim f_2(t_1, t_2) = \left[\frac{a\lambda_0}{(a + b)}\right]^2 - \frac{a\lambda_0 b\,(2a + b)\, e^{-(a+b)\tau}}{2\,(a + b)^2} \tag{23}$$

3. SHOT NOISE

For historical reasons, let us consider the voltage fluctuations in an anode circuit. In particular, if the anode circuit has an inductance L as well as a resistance R, and capacitance C, the voltage $V(t)$ due to the arrival of an electron at time $t = 0$ is given by (see Rowland[6])

$$V(t) = \frac{1}{2}\left(-\frac{\mathcal{E}}{c}\right)\left[\left(1 + \frac{R}{2i\omega L}\right) e^{-(\frac{R}{L} - i\omega)t} + \left(1 - \frac{R}{2i\omega L}\right) e^{-(\frac{R}{L} - i\omega)t}\right]$$
$$= 0 \qquad t < 0 \tag{24}$$

where ω is given by

$$\omega = \left(\frac{1}{CL} - \frac{R^2}{4L^2}\right)^{1/2} \tag{25}$$

Defining the Fourier transform of $V(t)$ as

$$w(\nu) = (2\pi)^{-1/2} \int_0^\infty V(t)\, e^{i\nu t}\, dt \tag{26}$$

we notice that the Fourier transform of the response function $\phi(t)$ is nothing but $W(\nu)\, G(\nu)$ where $G(\nu)$ is the amplification factor set for the frequency $\nu/2\pi$ cycles per second.

Thus, a study of $r(t)$ the cumulative response or rather the probability distribution of $r(t)$, leads us toward a good understanding of the physical phenomenon. The earliest attempts, therefore, were directed toward the calculation of the first two moments of $r(t)$. Campbell[8] proved that in the case of a stationary system, the mean and fluctuation about the mean of $r(t)$ are given by

$$\mathcal{E}\left\{r(t)\right\} = \lambda \int_0^\infty \phi(t')dt' \tag{27}$$

$$\mathcal{E}\left(\left[r(t) - \mathcal{E}\left\{r(t)\right\}\right]^2\right) = \lambda \int_0^\infty \left[\phi(t')\right]^2 dt' \tag{28}$$

where λ is the mean density of arrivals of electrons at the anode. Campbell assumed, as may be reasonably expected, that the number of electron arrivals at the anode is governed by a Poisson law which states that the probability that there are n arrivals in a time interval $(0, t)$ is $\pi(n, t)$ where

$$\pi(n, t) = e^{-\lambda t} \frac{(\lambda t)^n}{n!} \tag{29}$$

We observe that while (27) and (28) is true only for the stationary system, results can be generalized with equal facility to the non-stationary case (when t is finite) (see, for example, Ref. 9 and 10). However, (27) and (28) do not explain the shot effect completely when we take into account the limitations due to space-charge effect. In fact, Hull and Williams[11] made intensive measurements of the shot voltage soon after Campbell proposed his formulae and found that the measured shot voltage even fell below 40 percent of the theoretical value. Johnson[12] meanwhile, who made some measurements of shot voltage in space charge limited tubes, pointed out that the expected value of the shot voltage should be calculated for temperature limited currents by assuming an internal resistance of the valve. Based on these experimental findings, as well as the investigations of Moullin,[13] Rowland[14] formulated the problem in a precise form. He assumed that electrons with specific probabilities may have lives of any length in the anode system during which time each of them add $(-\epsilon/C)\exp[-(t-t_i)/RC]$ to the anode potential of the valve whose anode to earth capacitance is C and feed resistance is R. In addition, there is an effective internal resistance ρ of the valve defined by assuming that a variation of the anode potential causes a variation $1/\epsilon\rho$ times as great in the probable density of arrival of electrons. Thus, the arrival of each electron must decrease the probable density of further arrivals by an amount $(1/C\rho)\exp[-(t-t_i)/CR]$. If N_0 is the density of arrivals at time $t=0$, then $N(t)$, the arrival density at any time $t>0$, is given by

$$\begin{aligned} N(t) &= N_0 - \sum_i \alpha(t-t_i) \qquad && N_0 - \sum_i \alpha(t-t_i) > 0 \\ &= 0 && \text{otherwise} \end{aligned} \tag{30}$$

where

$$\alpha(t-t_i) = \left(\frac{1}{C\rho}\right)\exp\left[-\frac{(t-t_i)}{CR}\right] \tag{31}$$

With the modification of the density of arrivals as given by (30), it is clear that the probability that there are n arrivals in the interval $(0,t)$ is no longer given by a simple Poisson law as (29). The calculations of Rowland relating to the mean and mean square of the response were based on the behavior of $N(t)$ as given by (30). The expressions for the mean and mean square values of stationary $r(t)$ in terms of an infinite series of integrals under the exponential, though not formidable, were sufficiently complicated. These difficulties were in fact overcome by Rowland, who obtained simple expressions for the first two moments of stationary $r(t)$. However, there were some limitations in his final results. During the process of integration, he ignored the possibility of $N(t)$ dropping down to negative values and, therefore, there was no mechanism for achieving it in any refined form of calculation of the integrals. In spite of all this, Rowland's results can be regarded as a decisive achievement in the theory of shot-effect for inclusion of all the realistic features of shot noise. In fact, it is precisely for this reason that a considerable part of Moullin's book[15] is devoted to the presentation and discussion of Rowland's theory.

If $\phi(t)$ is the response to a single pulse, $\phi(t)$ is given by

$$\phi(t) = (2\pi)^{-1/2} \int_{-\infty}^{+\infty} W(x)\, G(x)\, e^{itx} dx \tag{32}$$

On the basis of (32) and (30), Rowland made the pioneer calculations for the mean-square response. However, the possibility of $N(t)$, the density of arrivals, taking negative values could not be circumvented. Rowland[14] first proved that

$$\lim_{\substack{t_1 \to \infty \\ t_2 \to \infty}} f_2(t_1, t_2)_{t_2 - t_1 = \tau} = \bar{N}^2 - \left(\frac{\bar{N}}{C\rho}\right) \exp\left[-\tau\left(\frac{1}{C\rho} + \frac{1}{CR}\right)\right] \tag{33}$$

and in a subsequent note pointed out that (33) is incorrect and should be replaced by

$$\lim f_2(t_1, t_2) = \bar{N}^2 - \left(\frac{2\bar{N}}{2\rho + R}\right) \exp\left[-\tau\left(\frac{1}{C\rho} + \frac{1}{CR}\right)\right] \tag{34}$$

This was a conjecture based on the experiments of Moullin. Equation (34) is identical with (23) if we replace a and b by $1/C\rho$ and $1/CR$, respectively. Thus, the results of the previous section fill up the gap anticipated by Rowland and provide rigorous proof of his results. Moreover, the method of approach explained in the non-Markovian process section, wherein we have averaged over ensembles making use of the fact that probability magnitudes can never be negative, removes all doubts regarding the correctness of the formula (34).

The results of the previous section enable us to obtain the correlation of the response function at two different times after the system has attained stationariness? Toward this end, we notice that if $N(t)$ represents the number of pulses up to time t, the cumulative response $r(t)$ is given by

$$r(t) = \int_0^t dN(\tau)\, \phi(t - \tau) \tag{35}$$

where $\phi(t)$ is given by (32). The correlation of $r(t)$ is given by

$$\mathcal{E}\left\{r(t_1)r(t_2)\right\} = \int_0^{t_1}\int_0^{t_2} \mathcal{E}\left\{dN(\tau_1)dN(\tau_2)\right\} \phi(t_1 - \tau_1)\, \phi(t_2 - \tau_2) \tag{36}$$

The integrand in (36) can be expressed in terms of the product densities of the pulses if we take into account the degeneracy arising from the overlapping of the intervals $d\tau_1$ and $d\tau_2$. Thus, we can use the general formula for the correlation obtained by Srinivasan and Vasudevan[3] for the treatment of Barkhausen noise:

$$\mathcal{E}\left\{r(t_1)\,r(t_2)\right\} = \int_0^{min(t_1,t_2)} f_1(\tau_1)\,\phi(t_1-\tau_1)\,\phi(t_2-\tau_1)d\tau_1$$

$$+\int_0^{t_1}\int_0^{t_2}\phi(t_1-\tau_1)\,\phi(t_2-\tau_2)\,f_2(\tau_1-\tau_2)\,d\tau_1 d\tau_2 \qquad (37)$$

We are interested in the limiting form of the right-hand side of equation (37) when t_1 and t_2 tend to infinity, while $t_2 - t_1$ remains fixed and is equal to C.

It is shown in Appendix B (of Ref. 3) that the Fourier transform of the left-hand side of (37), which is only a function of the single argument $t_2 - t_1$, can be expressed in terms of the Fourier transform or $\phi(t)$ and of the limiting form of $f_2(t_1, t_2)$ [see (34) which is again a function of the single argument $t_2 - t_1$]. Thus, $r(\omega)$, the power spectrum of the response* (which is nothing but the Fourier transform of the correlation), is given by

$$r(\omega) = (2\pi)^{1/2}\frac{a\lambda_0}{a+b}|\phi(\omega)|^2 + 2\pi\left(\frac{a\lambda_0}{a+b}\right)^2 \operatorname{Re} R(\omega)|\phi(\omega)|^2 \quad (38)$$

where $R(\omega)$ is the Fourier transform of $f_2(t_1, t_2)$.

4. BARKHAUSEN NOISE

The statistical theory of Barkhausen noise can be formulated in terms of a random process associated with correlated pulse trains. In view of the difficult nature of the problem arising from the dependence of probability of any event on an earlier event, not much attention has been paid, sofar, to the study of such processes. Very recently, Mazetti[17] studied the noise produced by correlated pulse trains by introducing a correlation between successive pulses. More specifically, he assumed that the time interval x between any two successive events is governed by a probability frequency function $P(x)$ where

$$P(x) = \nu^2 x \exp(-\nu x) \qquad (39)$$

Such a process is usually called a renewal process (see, for example Bartlett[5]). This is the simplest case of a non-Markovian process. However, in Barkhausen noise, the average number of pulses

* We use the same symbol ϕ to denote the function as well as its Fourier transform.

per unit time is not necessarily a constant, but depends on the instantaneous value of the macroscopic magnetization due to the turning of microscopic Weiss fields. To incorporate such effects, it is necessary to consider random processes in which ν is not only a function of t (t standing for the time parameter), but is dependent on the *random times* at which the events have occurred prior to t. In fact, the model proposed in Section 2 can be suitably modified so that the Weiss fields, as they get oriented by the external field, facilitate further turning of the microscopic magnets. Adopting the same line of approach, we shall assume that the impulse probability $\lambda(t)$ is random where the realized value at time t, corresponding to a set of n earlier events occurring at times $t_1, t_2, \ldots, t_n$ is given by

$$\lambda^R(t) = \lambda_0 + b \sum_{i=1}^{n} \exp\left[-a(t - t_i)\right] \tag{40}$$

This means that the occurrence of an event at time between t_i and $t_i + dt_i$ increases the probability of any later event at t by b [exp $-$ $a(t - t_i)$] We notice that unlike the case of shot noise where $\lambda(t)$ decreases due to the occurrence of events, in the present case, it increases with the events. Given that an event has occurred with probability $\lambda_1 dt_1$ at a time between t_1 and $t_1 + dt_1$, the probability that the next event occurs between t_2 and $t_2 + dt_2$ is given by

$$P(t_2|t_1)dt_2 = \exp\left[-\int_{t_1}^{t_2}\left(\lambda_1 + b\, e^{-a(t'-t_1)} + (\lambda_1 - \lambda_0)e^{-a(t'-t_1)}\right) dt'\right]$$
$$\times \left\{\lambda_0 + b\, e^{-a(t_2-t_1)}\right\} dt_2 \tag{41}$$

This may be called the conditional frequency function. This is the nearest equivalent of the function $P(x)$ dealt with by Mazetti. This is because unlike $P(x)$, the expression equation (41) is not a fully determined function of x since this function depends on the times of occurrence of earlier events which are themselves randomly distributed. We notice that we are interested in finding the moments and correlations of the responses $\psi(t)$ defined by

$$\psi(t) = \int_0^t \phi(t - \tau)\, dN(\tau) \tag{42}$$

where $N(\tau)$ represents the number of pulses up to time τ so that $dN(\tau)$ represents the number of pulses into the infinitesimal interval $(\tau, \tau + d\tau)$. To this end, we need the product densities $f_1(t_1)$ and $f_2(t_1, t_2)$ of events on the t-axis. This, in turn, can be obtained if we can calculate explicitly the moments of λ_1 at every point t_1 and the conditional moments of λ_2 at a point t_2 conditional upon λ having assumed a given value λ_1 at t_1. Since the calculations are similar to those exhibited in the non-Markovian process section, we refrain from giving the details which may be found in Ref. 3. The product densities in this case are given by

$$f_1(t) = \frac{a\lambda_0}{a-b} - \frac{b\lambda_0}{a-b} e^{-(a-b)t} \tag{43}$$

$$f_2(t_1, t_2) = p(2, t_1) e^{-(a-b)(t_2-t_1)} + p(1, t_1)\left[b\, e^{-(a-b)(t_2-t_1)} + \frac{a\lambda_0}{a-b}(1 - e^{-(a-b)(t_2-t_1)})\right] \tag{44}$$

When t_1 and $t_2 \to \infty$ it is seen that $f_2(t_1, t_2)$ is a function of $t_2 - t_1$ only:

$$\lim_{t_2-t_1=\tau} f_2(t_1, t_2) = \left(\frac{a\lambda_0}{a-b}\right)^2 + \frac{a\lambda_0 b}{2(a-b)^2}(2a-b)e^{-(a-b)\tau} \tag{45}$$

We may observe that the asymptotic form of $f_2(t_1, t_2)$ and $f_1(t_1)$ obtained in this section bear a great similarity to $f_2(t_1, t_2)$ and $f_1(t_1)$ corresponding to the interval distribution function $P(x)$ defined by Mazetti[17]. Using these functions we can calculate as before the mean or mean-square response and also the correlation function $\rho(b)$ as in the earlier section. The power spectrum of the response is given by

$$\rho(\omega) = 2\pi\nu\bar{a}^2 \operatorname{Re}|F(\omega)|^2 + 4\bar{a}^2 \operatorname{Re} 2\pi|F(\omega)|^2 R(\omega) \tag{46}$$

where $F(\omega)$ the Fourier transform of $F(\tau)$ is given by

$$F(\omega) = \frac{1}{2\pi}\frac{1}{\alpha - i\omega} \tag{47}$$

while

$$R(\omega) = \int_0^\infty R(\tau) e^{i\omega\tau}\, d\tau \tag{48}$$

Thus,

$$\rho(\omega) = \frac{a\lambda_0}{a-b}\frac{\bar{a}^2}{2\pi(\alpha^2+\omega^2)} + \bar{a}^2\left[\frac{2\delta(\omega)}{\alpha^2+\omega^2}\left(\frac{a\lambda_0}{a-b}\right)^2 + \frac{1}{\pi}\frac{ab\lambda_0}{(a-b)^2}\frac{(2a-b)(a-b)}{(a-b)^2+\omega^2}\right] \tag{49}$$

It is interesting that the shape of the power spectrum is of the same nature as given by Mazetti.[17] However, there is one important feature that makes this model distinct from a renewal stochastic process and which is due to the difference in the sign of the exponential term in the limiting form of the correlation functions.

5. CONCLUDING REMARKS

Finally, we wish to make a few general remarks on the applicability of the stochastic process described in Section 2 to other physical situations. A stochastic process of this type is provided by the photon correlations in the recent experiments of Hanbury-Brown and Twiss.[18] In the treatment of photon correlations, it is assumed that electron emission is essentially a Poisson process, with the Poisson parameter being governed by some probability distribution function (see, for example, Mandel,[19] Mandel, Wolf, and Sudarshan[20]). There is ample scope for improving the usual formula for intensity correlations on the basis of the results obtained in this paper. We believe also that some models of one-dimensional fluid binary mixtures (see, for example, Ref. 21) can be dealt with on the basis of the present results. Since the interparticle potential is of the exponential type, the process will very well fit in with our mode of description. However, in this case the problem is more difficult since we have to obtain higher order conditional correlation functions. Nevertheless, the present formulation can be used to obtain the second order correlation function and other results based on it. Apart from this, there are a number of questions regarding secondary emission in shot noise. The power spectrum of response function undergoes considerable modification, and we believe that a detailed study based on the present model may yield fruitful results.

REFERENCES

1 S. K. Srinivasan, *On a Class of Non-Markovian Processes*, I. I. T. preprint (1962).

2 S. K. Srinivasan, *Nuovo Cimento* **38**: 979 (1965).

3 S. K. Srinivasan and R. Vasudevan, *On a Class of Non-Markovian Processes Associated with Correlated Pulse Trains and Their Application to Barkhausen Noise*, I. I. T. Research Report No. 4 (1965).

4 L. Takacs, On secondary processes generated by a Poisson process and their applications in physics, *Acta Math. Acad. Sci. Hung.* **5**: 203–236 (1954).

[5] M. S. Bartlett, *Stochastic Processes*, Cambridge University Press (1955), p. 54.
[6] E. N. Rowland, The theory of shot effect 11, *Proc. Camb. Phil. Soc.* **33**: 344–358 (1937).
[7] A. Ramakrishnan, Stochastic processes relating to particles distributed in a continuous infinity of states, *Proc. Camb. Phil. Soc.* **46**: 595 (1950).
[8] N. Campbell, *Proc. Camb. Phil. Soc.* **15**: 117 (1909).
[9] J. E. Moyal, *J. Roy. Statist. Soc. B* **11**: 150 (1949).
[10] S. K. Srinivasan and P. M. Mathews, *Proc. Nat. Inst. Sci.* (*India*) **22** A: 369 (1956).
[11] A. W. Hull and N. H. Williams, *Phys. Rev.* **25**: 147 (1925).
[12] J. B. Johnson, *Phys. Rev.* **26**: 71 (1925).
[13] E. B. Moullins, *Proc. Roy. Soc.* **147** A: 100 (1934).
[14] E. N. Rowland, *Proc. Camb. Phil. Soc.* **33**: 344 (1937).
[15] E. B. Moullins, *Spontaneous Fluctuations of* Voltage, Clarendon Press, Oxford (1938).
[16] E. N. Rowland, *Proc. Camb. Phil. Soc.* **34**: 329 (1938).
[17] P. Mazetti, *Nuovo Cimento* **25**: 1322 (1962); **31**: 38 (1964).
[18] R. Hanbury-Brown and R. Q. Twiss, *Phil. Mag.* **45** : 663 (1954) ; *Proc. Roy. Soc.* (*London*) **242** A: 300 (1957); **243** A: 291 (1957).
[19] L. Mandel, *Proc. Phys. Soc.* **72**: 1037 (1958).
[20] L. Mandel, E. C. G. Sudarshan, and E. Wolf, *Proc. Phys. Soc.* **84** : 435 (1964).
[21] R. Kikuchi, *J. Chem. Phys.* **23**: (1955).

Many-Particle Structure of Green's Functions

K. SYMANZIK*

NEW YORK UNIVERSITY
New York, New York

Since the reader may be unfamiliar with many-particle structure analysis (MPSA) of Green's functions, I shall first present an introduction to this subject. The setting of MPSA is "axiomatic quantum field theory." Axiomatic field theory, however, has many aspects; I will list, therefore, the various approaches, point out their distinctive features, and determine the relative position of MPSA. (The term "axioms" as used here merely denotes tentative postulates of a theory that it is hoped will describe elementary particle physics.) The following axioms common to all approaches (with one exception) are too well known to need discussion here (see, for example, Streater and Wightman, and Schweber[1]):

I. Invariance under the inhomogeneous proper Lorentz group.
II. Existence of local field operators.
III. Mass spectrum conditions.
IV. Completeness of the asymptotic particle states.

We may now classify these approaches according to rigor, that is, according to the generality and precision of formulation of the axioms as well as the mathematical purity of the deductions drawn from them on their respective levels (see Fig. 1).

As I have diagrammed them, among A to D, the lower-lying levels are less "rigorous" than higher-lying ones and, therefore, are specializations of the latter.

---

* Department of Mathematics.

A. Wightman approach: only axioms I–IV	⟵ ? ⟶	E. Haag–Araki approach: II replaced by concept of local operator rings
B. LSZ approach: assumption of existence and certain properties of Green's functions		
C. Derivation of analytic properties of scattering amplitudes under additional technical assumptions (Bogoliubov, Lehmann, Mandelstam)	⟵ D. MPSA of Green's functions, involves technical assumptions	
F. Lagrangian quantum field theory	G. Landau–Chew approach: analyticity assumptions about scattering amplitudes	

A few comments: The completeness axiom IV, originally stated only on the B level, can be stated now also for A, owing to the Haag–Ruelle scattering theory[2] which is based essentially on axioms I–III only. For the Green functions of the B levels, we shall go into details later. The technical assumptions on the C level will not be given here. Recently Hepp[3] has shown that, for example, for the pion–nucleon elastic forword scattering amplitude, the dispersion relation can be already derived on the A level. Bros, Epstein, and Glaser[4] have shown that some results of the kind Mandelstam[5] has obtained (analyticity in energy and momentum transfer simultaneously) can, in some domain smaller than that implied by the Mandelstam representation, already be obtained on the B level. This also holds, for say, nucleon–nucleon scattering, for which no analyticity in energy could be proved with methods used previously. In D, not only are technical assumptions made as well different from those under C, but the aim is different—although this goal, when achieved by sufficient work, should help C also, as I have indicated by the arrow and shall explain more specifically later. Moreover, I expect that by a sufficiently penetrating analysis, which could justify the so-far "technical" assumptions, MPSA could be raised to the B level. (Certain simple parts of D have, in fact, already been so raised.[6])

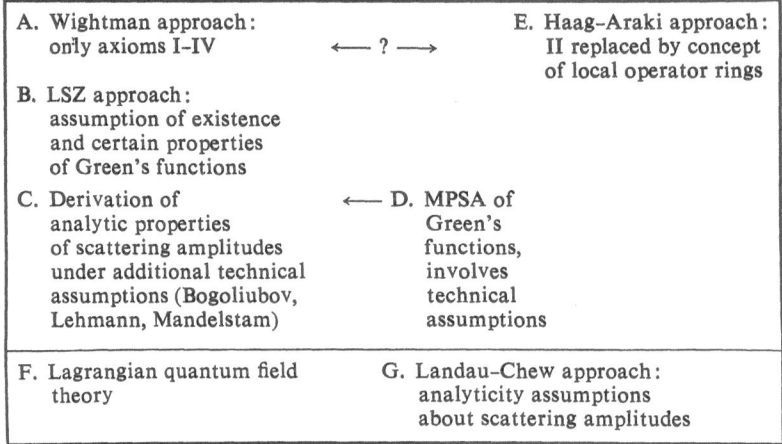

The E approach is currently undergoing such extensive development that I cannot say what its present results are compared with A–C. Vigorous attempts are being made now from both directions to unify A and E as far as possible. (Incidentally, Haag is responsible also for part of the B approach, while my own contribution to B is peripheral.)

D is clearly related to an entirely different approach to elementary particle physics, namely, Lagrangian quantum field theory (F); I shall elaborate on this later. There is also an as yet rather intuitive connection to analytic *S*-matrix theory (G) that raises analyticity assumptions about scattering amplitudes, to "axioms" by requiring that the analyticity be as near the perturbation theoretical one as is consistent with unitarity and existence.

After this review of axiomatism, I feel I should justify my concern with such a low-lying subject as MPSA. I have already stressed that all "axioms" mentioned are tentative postulates that mark a frame in which we are searching for a theory describing elementary particle physics. As long as it is not known whether these postulates admit a solution with a nontrivial *S*-matrix (not to speak of the "true" one), and as long as it is not certain whether such a basic phenomenon as general relativity can be described within this frame, I feel that a very lenient attitude toward "axioms" is in order and that the old LSZ maxim of putting interest of results above rigor of their derivation is still appropriate to the situation. (In a field where little except the free field is known for certain, taste reigns supreme as arbiter, and tastes always differ.)

It is sometimes suggested that the axiomatic frame is too general for imposing locality only, while the "dynamics" is merely hoped to be found later inside that frame after an entirely abstract analysis. While this may or may not be a valid objection, at present, "technical assumptions" serve to narrow the frame (if they do so at all), and if these assumptions allow us to establish a relation with, e.g., Lagrangian quantum field theory they deserve to be made and followed up. Of course, it also might be interesting to investigate what conclusions can be reached if no technical assumptions are made, but it is difficult to do too many things at the same time. MPSA appeared to me to be the almost logical extension of the LSZ approach.

In character, MPSA is, so far (at least in its less trivial parts), a formal approach. No approximate integral equations are being proposed or solved, and whenever an assumption on, e.g., the solvability of a nonapproximate, integral equation is needed, it is made even on insufficient mathematical evidence. (Clearly, once the formal structure of certain objects is understood, it will be easier to find approximation schemes and, especially, to speculate on their meaningfulness.) The guideline of the formal analysis is the structure of perturbation theory. The value of perturbation theory is today much more appreciated than it was, say, five years ago. Since Landau's forceful argument[8] to take perturbation-theoretical analytic properties simply as "valid" ones, considerable effort has been devoted to the study of Feynman graphs. But whereas, this study, so far lacks a connection with well-established theoretical high-energy physics (namely, quantum field theory), the use of perturbation expansions in MPSA is entirely different and in fact, could be simply omitted. One need merely take from Feynman (or other) graphs the idea as to how to proceed (and what technical assumptions to make); the procedure itself then turns out to be surprisingly successful and sheds new light on all nonperturbation-theoretical approaches that use the Bethe-Salpeter equation or one of its descendants and which have persevered through the years. This also answers the question: How does it come about that perturbation expansions formally solve the basic integral equations for Green's functions on the LSZ level? I will now turn to these equations and to the heart matter of our subject.

In doing so, I refer to my own work,[9–13] and to work by Wraith[14,15] and Streater.[6]

1. NOTATION

We shall be concerned with a theory of only one kind of neutral spinless particles of mass m and, therefore, need consider only one Hermitian scalar local field $A(x)$. The extension of all results to more general cases is straightforward; in fact, Wraith[14] has discussed such generalizations. I will only mention that, as Zimmermann[16] has shown, in the LSZ approach it does not matter if particles are "elementary" or "composite," whatever the distinction. Only their stability matters and makes it possible to associate with them a local field satisfying the asymptotic condition to be stated.

Explicitly, then, with the metric

$$g_{\mu\nu} = g^{\mu\nu} = \begin{pmatrix} 1 & & & \\ & -1 & & \\ & & -1 & \\ & & & -1 \end{pmatrix}$$

we have

$$[A(x), A(y)] = 0 \qquad \text{if} \qquad (x-y)^2 < 0$$

I will now briefly introduce the notation and definitions needed for our case:

$$\overleftrightarrow{\partial}_\mu = \frac{\overleftrightarrow{\partial}}{\partial x^\mu} = -\overleftarrow{\partial}_\mu + \overrightarrow{\partial}_\mu$$

the arrow pointing to the term to be differentiated,

$$K = \partial_\mu \partial^\mu + m^2$$

Green's theorem

$$\oint_{\partial\Omega} f \overleftrightarrow{\partial}_\mu g \, d\sigma^\mu = \int_\Omega f(-\overleftarrow{K} + \overrightarrow{K}) g \, dx \tag{1}$$

(Here, Ω is a domain in four-dimensional space–time, $\partial\Omega$ is its boundary, $d\sigma^\mu$ is the surface element with normal pointing outward, and

$$d\sigma^\mu = \varepsilon^{\mu\nu\kappa\lambda} \delta_{a\nu} \delta_{b\kappa} \delta_{c\lambda} \qquad \varepsilon^{0123} = -1$$

where δ_a, δ_b, δ_c span a parallelepiped in the right-hand sense. Actually, the use of general instead of flat surfaces is only in our context, a matter of esthetics.)

The fundamental solution of the homogeneous Klein–Gordon equation

$$\Delta(x-y) = \frac{-i}{(2\pi)^3} \int dk e^{-ikx} \delta(k^2 - m^2)\, \varepsilon(k_0)$$

[$\varepsilon(k_0)$ = sign k_0 should in manifestly covariant expressions multiply only functions of k that vanish for $k^2 < 0$] satisfies

$$K\Delta(x-y) = 0 \qquad \Delta(x-y)_{x^0=y^0} = 0$$

$$\partial_{x^0} \Delta(x-y)_{x^0=y^0} = -\delta(\vec{x} - \vec{y})$$

and if $Kf = 0$, then

$$f(x) = -\int \Delta(x-y) \overleftrightarrow{\partial}_{\mu y} f(y)\, d\sigma_y^\mu \tag{2}$$

independently of the choice of the (globally spacelike) surface. We also need

$$\pm i\Delta^\pm(x-y) = \frac{1}{(2\pi)^3} \int dk\, e^{-ikx} \delta(k^2 - m^2)\, \theta(\pm k_0)$$

with $\theta(k_0) = [1 + \varepsilon(k_0^+)]/2$ to be used similarly as $\varepsilon(k_0)$. For $f(x)$

with $Kf(x) = 0$ we write

$$f(x) = \frac{1}{(2\pi)^{3/2}} \int dk\, \delta(k^2 - m^2) \tilde{f}(k)\, e^{-ikx} = f^+(x) + f^-(x)$$

with

$$f^{\pm}(x) = - \int \Delta^{(\pm)}(x - y) \overleftrightarrow{\partial}_{\mu y} f(y)\, d\sigma_y^\mu \tag{3}$$

and define for a pair f, g with $Kf = 0$, $Kg = 0$

$$(\bar{f}, g) = i \int \bar{f}\, \overleftrightarrow{\partial}_\mu g\, d\sigma^\mu = \int dk\, \bar{\tilde{f}}(k)\, \delta(k^2 - m^2)\, \tilde{g}(k)\, \varepsilon(k_0) \tag{4}$$

independent of the choice of the (globally spacelike) surface. A complete orthonormal set of positive-frequency solutions (f_α), $f_\alpha^+ = f_\alpha, f_\alpha^- = 0$ is defined by

$$\sum_\alpha f_\alpha(x)\, \bar{f}_\alpha(y) = i\Delta^{(+)}(x - y) \qquad (\bar{f}_\alpha, f_\beta) = \delta_{\alpha\beta} \tag{5}$$

We need the functions

$$\Delta_{\substack{\mathrm{ret} \\ \mathrm{av}}}(x) = \mp\theta(\pm x^0)\Delta(x) = \frac{1}{(2\pi)^4} \int \frac{e^{-ikx}}{m^2 - k^2 \mp i\varepsilon k^0} dk \qquad \varepsilon \to +0 \tag{6}$$

and

$$\begin{aligned} \Delta_F(x) &= -i\Delta_{\mathrm{ret}}(x) - i\Delta^{(-)}(x) \\ &= -i\Delta_{\mathrm{av}}(x) + i\Delta^{(+)}(x) \\ &= \frac{-i}{(2\pi)^4} \int \frac{e^{-ikx}}{m^2 - k^2 - i\varepsilon} dk \qquad \varepsilon \to +0 \end{aligned} \tag{7}$$

which satisfy

$$K\Delta_{\mathrm{ret}}(x) = K\Delta_{\mathrm{av}}(x) = iK\Delta_F(x) = \delta(x) \tag{8}$$

2. FUNCTIONALS

We now introduce the tools that make MPSA possible in a reasonably compact form, namely, functionals.

A functional[17] associates with a function, e.g., $j(x)$ with $x \epsilon R$, a number (or function of some parameters, or operator) $F\{j\}$, i.e., mathematically represents a mapping of a vector into a scalar (or vector). $F\{j + zj'\}$ is a functional of $j(x)$ and $j'(x)$ but an ordinary function of z, for which the Taylor formula holds:

$$F\{j + zj'\} = F\{j\} + z\frac{\partial}{\partial z}F\{j + zj'\}|_{z=0}$$

$$+\cdots+\frac{z^n}{n!}\frac{\partial^n}{\partial z^n}F\{j+zj'\}|_{z=0}$$

$$+\int_0^z dz'\frac{(z-z')^n}{n!}\frac{\partial^{n+1}}{\partial z'^{n+1}}F\{j + z'j'\} \qquad (9)$$

This may be written differently by use of functional derivatives

$$\frac{\partial F\{j\}}{\partial j(y)} = \lim_{\varepsilon\to 0}\frac{F\{\hat{j}\} - F\{j\}}{\varepsilon}$$

with $\hat{j}(x) = j(x) + \varepsilon\delta(x - y)$ or, more generally,

$$\int dyj'(y)\frac{\delta F\{j\}}{\delta j(y)} = \lim_{\varepsilon\to 0}\frac{F\{j + \varepsilon j'\} - F\{j\}}{\varepsilon} \qquad (10)$$

Then, the $(l + 1)^{\text{th}}$ term $(l \leqslant n)$ on the right-hand side of (9) becomes, with iterated use of this definition,

$$\frac{z^l}{l!}\int\cdots\int \mathrm{d}y_1\ldots dy_l j'(y_1)\ldots j'(y_l)\frac{\delta^l F\{j\}}{\delta j(y_1)\ldots\delta j(y_l)}$$

with

$$\frac{\delta^l F\{i\}}{\delta j(y_1)\ldots\delta j(y_l)} = \frac{\delta}{\delta j(y_l)}\left(\frac{\delta^{l-1}F\{j\}}{\delta j(y_1)\ldots\delta j(y_{l-1})}\right)$$

$$= F_{y_1}\ldots_{y_l}\{j\}$$

Similarly as for partial derivatives,

$$F_{y_1}\ldots_{y_l}\{j\} = F_{y_{i_1}}\ldots_{y_{i_l}}\{j\}$$

$(i_1\ldots i_l$ is a permutation of $1\ldots l)$ if the derivative exists and is continuous.

Letting $n\to\infty$ in (9), we obtain the Volterra series

$$F\{j + zj'\} = F\{j\} + \sum_{n=1}^{\infty}\frac{z^n}{n!}\int\cdots\int dy_1\ldots dy_n\cdot j'(y_1)\ldots j'(y_n)F_{y_1}\ldots_{y_n}\{j\}$$

$$= \exp\,[zj'(\delta/\delta j)]F\{j\} \qquad (11)$$

with $j'(\delta/\delta j) = \int dxj'(x)(\delta/\delta j(x))$, which is the generalization to an infinite number of dimensions of the elementary formula

$$f(x_1 + \alpha_1,\ldots, x_n + \alpha_n) = \exp\left[\sum_{\nu=1}^{n}\alpha_\nu\frac{\partial}{\partial x_\nu}\right]f(x_1,\ldots, x_n)$$

As (11) stands, it involves a convergence assumption (i. e., that the remainder term in (9) goes to zero as $n\to\infty$). However, if, e. g., we set $j = 0$, $z = 1$, and $F_{y_1\ldots y_n}\{0\} = F(y_1,\ldots, y_n)$, then we may consider

$F\{j'\}$ as formally defined by its Volterra expansion, i. e., as the collection of the number F{o} and all the symmetric functions

$$F(y_1, \ldots, y_n) = \frac{\delta^n}{\delta j(y_1) \ldots \delta j(y_n)} F\{j\}\Big|_{j=0} \tag{12}$$

as one often considers, e. g., formal power series without asking the convergence question. Operations on the expanded "function" become operations on the infinitely many coefficients, as is familiar to the reader from the use of generating functions in the theory of special functions of mathematical physics. Likewise, we may consider operations (e.g., functional differentiations) on functionals as operations on the expansion coefficients. Thus if, as in (12), $F\{j\}$ is the generating functional of the infinite set of functions $F(y_1, \ldots, y_n)$ with $n = 0, 1, 2, \ldots$, then

$$\frac{\delta^l F\{j\}}{\delta j(x_1) \ldots \delta j(x_l)}$$

is the generating functional of the set of functions $F(x_1, \ldots, x_l, y_1, \ldots, y_n)$.

Note that expandability of proper (i.e., nonformal) functions is a very special property. For example, the functional $F\{j\} = \max |j(x)|$ with $f \varepsilon C^0$ does not have it [in fact, does not even possess a functional derivative as a linear functional of j' in the sense of (10)].

Occasionally, we may think of a functional as a proper one. The following example, in fact, comes very close to our case: Let a string be stretched along the x axis between $x = 0$ and $x = l$. A unit weight at y will cause at x the deplacement $G(x, y)$. Assuming linearity (by Hooke's law valid for small displacements) the weight distribution $\rho(y)dy$ will give at x the displacement $F_x\{\rho\} = \int G(x,y)\rho(y)dy$. Due to the well-known symmetry of the Green's function (reciprocity principle)

$$\frac{\delta}{\delta\rho(y)} F_x\{\rho\} = G(x, y) = G(y, x) = \frac{\delta}{\delta\rho(x)} F_y\{\rho\} \tag{13}$$

$F_x\{\rho\}$ may be written

$$F_x\{\rho\} = \frac{\delta}{\delta\rho(x)} F\{j\}$$

with $F\{j\} = \frac{1}{2} \int\int dx\, dy\, j(x) j(y) G(x, y)$, the negative of the potential energy. The equality of the first and last term in (13) is the "integrability condition." The quantum-mechanical analog of the "influence functional" $F\{j\}$ are just the functionals we will be dealing with. (For elaboration and further examples see Ref. 11 or any book on nonlinear continuum mechanics or nonlinear integral equations.)

3. THE FREE FIELD

We need consider only a scalar Hermitian free field $A_0(x)$. It satisfies

$$KA_0(x) = 0 \qquad [A_0(x), A_0(y)] = i\Delta(x - y)$$

Then, as in (4),

$$(A_0, f) = i \int A_0 \overleftrightarrow{\partial}_\mu f \, d\sigma^\mu \equiv A_0^f \tag{14}$$

and as in (3)

$$A_0(x) = A_0^+(x) + A_0^-(x)$$

The vacuum is defined by

$$A_0^+(x)\rangle = 0 \qquad \text{for all } x \tag{15}$$

We have (Exercise)

$$\langle A_0(x), A_0^f\rangle = f^+(x)$$

$$\langle A_0^f, A_0(x)\rangle = -f^-(x)$$

and for the time-ordered product

$$\langle TA_0(x)\, A_0(x)\rangle = \Delta_F(x - y)$$

$$= \langle A_0(y)\, A_0(x)\rangle + \theta(x^0 - y^0)\langle[A_0(x), A_0(y)]\rangle$$

The factor of $\theta(x^0 - y^0)$ vanishes for $(x - y)^2 < 0$ but behaves singularly at $x \to y$ from time-like instead of space-like directions. This is the same situation as in (6) and (7), however, and may be handled by considering $\Delta_F(x - y)$ as a distribution to be used only with sufficiently smooth test functions. Systematically, we may introduce the generating functional of time-ordered products $TA_0(x_1) \ldots A_0(x_n)$ (time increasing from right to left),

$$T \exp\left[i\int j(x) A_0(x)\, dx\right]$$

Using

$$e^{-\lambda q} p e^{\lambda q} = p + \frac{\lambda}{1!}[p, q] + \frac{\lambda^2}{2!}[[p, q], q] + \ldots \tag{16}$$

(verified by differentiation with respect to λ) and, therefrom,

$$e^p e^q = e^q e^p e^{[p,q]} \qquad \text{if } [[p, q], p] = [[p, q], q] = 0 \tag{17}$$

It is easy to verify (e. g., by functional differentiation) that

$$T \exp\left[i \int j(x)A_0(x)\,dx\right] = :\exp\left[i \int j(x)A_0(x)\,dx\right]:$$
$$\exp\left[-\frac{1}{2}\iint dx\,dy\,j(x)j(y)\Delta_F(x-y)\right]$$

the closed form of Wick's theorem, with

$$:\exp\left[i\int j(x)A_0(x)dx\right]: = \exp\left[i\int j(x)A_0^-(x)dx\right]\exp\left[i\int j(x)A_0^+(x)dx\right]$$

the "normal" ordering. The unit operator in Hilbert space, spanned by applying operators $(A_0, f_\alpha) \equiv A_0^\alpha$, with f_α the complete orthonormal set of positive-frequency solutions of the Klein-Gordon equation considered in I, on the vacuum, is

$$1 = e^{A_0^-(\delta/\delta j)}\rangle\langle e^{(j, A_0^+)}|_{j=0} \tag{18}$$

(Here j is, effectively, a solution of the Klein-Gordon equation since it becomes $A^{(-)}$ and is introduced only to mark the position of the $A^{(\pm)}$ operators, respectiively, to the right and left of the projector on the vacuum. [Exercise: verify $\langle 1\rangle = 1$ and $[A_0(x), 1] = 0$ which proves (18).] Using $O = 1O1$, any operator in the Hilbert space can with (18) and (14) be manipulated into

$$O = :e^{A_0(\delta/\delta j)}: \langle e^{(j, A_0)} O e^{-(j, A_0)}\rangle|_{j=0}$$
$$= \langle O\rangle + \sum_{n=1}^{\infty}\frac{(-i)^n}{n!}\int\ldots\int d\sigma_{y_1}^{\mu_1}\ldots d\sigma_{y_n}^{\mu_n} A_0(y_1)\cdots A_0(y_n)$$
$$\times \overleftrightarrow{\partial}_{y_1\mu_1}\cdots\overleftrightarrow{\partial}_{y_n\mu_n}\langle[[[O, A_0(y_1)], A_0(y_2)]\cdots A_0(y_n)]\rangle \tag{19}$$

[Exercise: verify (19). Hint: Use (11) repeatedly, and under the condition of (17) $e^p e^q = e^{p+q+1/2[p,q]}$ proved similarly as (16).]

4. INTERACTING FIELD

Now let $KA(x) \neq 0$, but

$$[A(x), A(y)] = 0 \qquad \text{if } (x-y)^2 < 0 \tag{20}$$

(We use the Heisenberg representation throughout.) (A, f) defined analogously to (4), is now not surface-independent since only $Kf = 0$. For definiteness, we use a "flat" surface and introduce

$$A^f(t) = i\int_{x^0=t} A(x)\overleftrightarrow{\partial}_{x_0} f(x)\, d\vec{x}$$

We define (loosely)

$$A_{\substack{\text{in}\\ \text{out}}}(x) = A(x) - \int \Delta_{\substack{\text{ret}\\ \text{av}}}(x-y)\vec{K}A(y)\, dy \tag{21}$$

Then[16] (after, if necessary, changing A by a factor) $A_{\text{in}}(x)$ and $A_{\text{out}}(x)$ are free fields in the sense of III, with (due to the assumed uniqueness of the vacuum, axiom III) the same "true" vacuum$\rangle$. Due to axiom IV, the Hilbert spaces spanned by applying either the $(A_{\text{in}}, f_\alpha)$ or the $(A_{\text{out}}, f_\alpha)$ on $\rangle$ are the same Hilbert space of the theory, and the transformation of basis vectors is described by the unitary scattering operators S

$$A_{\text{out}}(x) = S^+ A_{\text{in}}(x) S \qquad S\rangle = S^+\rangle = \rangle \tag{22}$$

More correctly, (21) should be replaced by the LSZ asymptotic condition,[18]

$$\underset{t\to\mp\infty}{\text{weak limit}}\; A^{f_\alpha}(t) = (A_{\substack{\text{in}\\ \text{out}}}, f_\alpha) \qquad \text{for all } \alpha \tag{23}$$

The formulas (18) and (19) remain valid with A_0 read as A_{in} or A_{out}.

5. *T* AND *R* PRODUCTS. τ AND *r*-FUNCTIONS

We define the generating functional of time-ordered products

$$\begin{aligned} T\{j\} &= T\exp\left[i\int dx\, j(x)A(x)\right] \\ &= 1 + \sum_{n=1}^{\infty} \frac{i^n}{n!}\int\cdots\int dx_1\cdots dx_n\, j(x_1)\cdots j(x_n) \\ &\qquad\qquad \times TA(x_1)\cdots A(x_n) \end{aligned} \tag{24}$$

which is [as for the free field, see section (III)] independent of the choice of the time axis due to (20). I shall omit $\{j\}$ whenever no confusion may arise. Then,

$$\begin{aligned} T_x = \frac{\delta T\{j\}}{\delta j(x)} &= iA(x) + \sum_{n=1}^{\infty}\frac{i^{n+1}}{n!}\int\cdots\int dx_1\cdots dx_n\, j(x_1)\cdots j(x_n) \\ &\qquad\qquad \times T[A(x)A(x_1)\cdots A(x_n)] \end{aligned}$$

$$= T\exp\left[i\int_{x^{0\prime}=x^0}^{\infty} dx'j(x')A(x')\right]iA(x)$$

$$\times\ T\exp\left[i\int_{-\infty}^{x^{0\prime}=x^0} dx'j(x')A(x')\right] \qquad (25)$$

Partial integration, using (1), (23), (8), (7), and (5) gives

$$-\,iT_x = A_{\text{out}}^-(x)T\{j\} + T\{j\}A_{\text{in}}^+(x) + \int \Delta_F(x-y)\vec{K}T_y dy \qquad (26)$$

and likewise, from (1), (23), (2), and (22) we obtain

$$\left[A_{\text{in}}(x), ST\{j\}\right] = iS\int \Delta(x-y)\vec{K}T_y\, dy \qquad (27)$$

Iterating either (26) or (27) shows that integrations as in (26) as also in (27) are interchangeable among themselves, since

$$[A_{\text{in}}^+, A_{\text{in}}^+] = [A_{\text{out}}^-, A_{\text{out}}^-] = 0 \qquad \text{and } [A_{\text{in}}, A_{\text{in}}] = C\text{ number}.$$

Inserting (27) and all its iterations into (19), and using (2) gives with $S\rangle = \rangle$

$$ST\{j\} = :e^{A_{\text{in}}\vec{K}(\delta/\delta j)}: \langle T\{j\}\rangle \qquad (28)$$

the closed form of "reduction formulas." We call $\langle TA(x_1)\cdots A(x_n)\rangle = \tau(x_1\cdots x_n)$ a Feynman amplitude or τ-function. Equation (28) states that, for example, to calculate a matrix element of the S-operator, we need only know the singularity of a τ-function "on the mass shell." That the residuum of the singularity, the more precise nature of which is given by (26) or, by iteration,

$$\tau(x_1\cdots x_n) = i^n\int\cdots\int dy_1\cdots dy_n\Delta_F(x_1-y_1)\cdots$$

$$\Delta_F(x_n-y_n)\vec{K}_{y_1}\cdots\vec{K}_{y_n}\tau(y_1\cdots y_n) \qquad (29)$$

in general, does not vanish follows from (27). It is not difficult to show [e.g., by approximating (24) by a product similar to that as in (25) or, more rigorously, by considering a family of $T\{j\}$ with varying upper and lower bound in the integrations in (24)] that $T\{j\}$ is [for real $j(x)$] unitary:

$$T\{j\}^+T\{j\} = T\{j\}T\{j\}^+ = 1 \qquad (30)$$

Iterating (25) gives with (30)

$$T_{xy} = \theta(x^0-y^0)T_xT^+T_y + \theta(y^0-x^0)T_yT^+T_x$$

or,

$$- T^{+}T_{xy} = \theta(x^0 - y^0)T_x^{+}T_y + \theta(y^0 - x^0)T_y^{+}T_x \tag{31}$$

The Hermitian part of this gives essentially (30) while the anti-Hermitian part gives

$$- T^{+}T_{xy} + T_{xy}^{+}T = \varepsilon(x^0 - y^0)[T_x^{+}T_y - T_y^{+}T_x] \tag{32}$$

Inserting here (28) and using (17) gives for the vacuum expectation value of (32) an infinite system of coupled nonlinear integral equations for τ-functions. Any solution of that system and of that arising similarly from (30) gives, due to (28) a solution of axiomatic field theory [cf.[18] where (30) and (32) are replaced by a slightly different set of equations].

No such solution with a nontrivial S [to be obtained from (28)] is known. Therefore, we should try to extract partial information from (30) and (32). However, the infinite system of equations arising from (32) is difficult to analyze. Considerable simplification is achieved by introducing R-products and r-functions.[19]

Define the Hermitian operator-valued functional

$$R_x\{j\} = - iT\{j\}^{+}T_x\{j\} = iT_x\{j\}^{+}T\{j\} \tag{33}$$

from (30). Its Volterra expansion is

$$R_x\{j\} = A(x) + \sum_{n=1}^{\infty} \frac{1}{n!} \int \cdots \int dy_1 \cdots dy_n \cdot R(x, y_1 \cdots y_n) j(y_1) \cdots j(y_n) \tag{34a}$$

where [cf. (16)]

$$\begin{aligned} R(x, y_1 \cdots y_n) &= \frac{\delta^n}{\delta j(y_1) \cdots \delta j(y_n)} R_x\{j\}|_{j=0} \\ &= i^n \sum_{\text{perm}} \theta(x^0 - y_{i_1}^0)\theta(y_{i_1}^0 - y_{i_2}^0) \cdots \\ &\qquad \theta(y_{i_{n-1}}^0 - y_{i_n}^0)[\cdots[A(x), A(y_{i_1})] \cdots A(y_{i_n})] \end{aligned} \tag{34b}$$

are R-products or retarded multiple commutators. Since $T\{j\}$ is independent of the choice of coordinate system, due to (33), so also is $R_x\{j\}$. The vacuum expectation values of R-products, the r-functions

$$\langle R(x, y_1 \cdots y_n) \rangle = r(x, y_1 \cdots y_n)$$

are real invariant functions, symmetric in $y_1 \cdots y_n$, and vanish unless all y_i's are time-like (or light-like) earlier than x. From (31) and (33) follows

$$\frac{\delta}{\delta j(y)} R_x\{j\} = R_{x,y} = i\theta(x^0 - y^0)[R_x, R_y] \tag{35}$$

which may be separated into:

$$R_{x,y} = 0 \qquad \text{unless } y \text{ is time- or light-like earlier than } x \tag{36a}$$

and

$$R_{x,y} - R_{y,x} = i[R_x, R_y] \tag{36b}$$

From (28) and (33) follows, with (17)

$$R_x\{j\} = :e^{A_{\text{in}}\vec{K}(\delta/\delta j)}: \langle R_x\{j\}\rangle \tag{37}$$

which can also be derived by verifying (Exercise!)

$$[A_{\text{in}}(y), R_x] = i\int \Delta(y - y')\vec{K}_{y'} R_{x,y'} dy' \tag{38}$$

and using (19). Equation (37) allows us to consider in (36) the vacuum expectation value only,

$$\langle R_{x,y}\rangle = 0 \qquad \text{unless } y \text{ is time- or light-like earlier than } x \tag{39a}$$

and

$$\langle R_{x,y}\rangle - \langle R_{y,x}\rangle = i\langle R_x\rangle \left\langle \exp\left[\frac{\overleftarrow{\delta}}{\delta j} \overleftarrow{K} i\Delta^{(+)} \vec{K} \frac{\vec{\delta}}{\delta j}\right] - \exp\left[\frac{\overleftarrow{\delta}}{\delta j} \overleftarrow{K}(-i\Delta^{(-)}) \vec{K} \frac{\vec{\delta}}{\delta j}\right]\right\rangle \langle R_y\rangle \tag{39b}$$

where again (17) is used, since application of $\exp[A_{\text{in}}\vec{K}(\delta/\delta j)]$ on (39a and b) gives back (36). Equation (39a), if expressed in terms of r-functions, gives their retardedness, while (39b) is an infinite system of coupled nonlinear integral equations for these functions. It is simple, but suppressed here for brevity, to show[20] that finding a system of invariant real r-functions satisfying the conditions implied by (39) gives, via (37) and $R_x\{0\} = A(x)$, an operator field that satisfies the axioms I–IV. Again, no solution of physical interest is known. However, (39b) is much easier to analyze than the systems of equations derived from (30) and (32), as we shall later see in an example. [Exercise: write down the general r-function equation corresponding to (39b). Note that if Fourier transforms with respect to all variables are taken, for finite momenta, only a finite number of terms contribute in the equation.]

First, we note a few simple formulas:

$$R_x = A_{\text{in}}(x) + \int dx' \Delta_{\text{ret}}(x - x')\vec{K}R_{x'} \tag{40}$$

verified by partial integration and using (34b) and the asymptotic condition (23), and from (34b) or (35) we obtain

$$R_{x,y} = \int dy' \Delta_{\text{av}}(y - y')\vec{K}_{y'} R_{x,y'} \tag{41}$$

The ensuing formula analogous to (29) is

$$\begin{aligned} r(x, y_1 \cdots y_n) = \int dx' \int \cdots \int dy'_1 \cdots dy'_n \\ \times \Delta_{\text{ret}}(x - x')\Delta_{\text{av}}(y_1 - y'_1) \cdots \Delta_{\text{av}}(y_n - y'_n) \\ \times \vec{K}_{x'} \vec{K}_{y'_1} \cdots \vec{K}_{y'_n} r(x', y'_1 \cdots y'_n) \end{aligned} \tag{42}$$

which exhibits the nature of the singularities "on the mass shell," namely, retarded and advanced ones, of course. That the residua are again, generally, not zero follows from (38) and (Exercise!)

$$T^{+}\{j\}A_{\text{out}}(x)T\{j\} - A_{\text{in}}(x) = -\int dx' \Delta(x - x')\vec{K}R_{x'} \tag{43}$$

However, the Fourier transforms of the r-functions also possess other singularities than those shown in (42). To locate these and determine their nature is the aim of MPSA. In this, perturbation theory will be a guide, but not an instrument of proof; proofs should be based solely on axiomatic properties, e.g., the infinite system of equations (39b) and, if necessary, plausible "technical assumptions."

6. PERTURBATION SOLUTIONS

Assume the Lagrangian to be

$$L = \frac{1}{2} \partial^{\mu} A_u \partial_{\mu} A_u - \frac{m_u^2}{2} A_u^2 - H_i(A_u) \tag{44}$$

with $H_i(A_u)$ a polynomial. The subscript u denotes "unrenormalized" quantities, i.e., $m_u^2 \neq m^2$ (the particle mass) and A_u only proportional to A, the field operator that satisfies the asymptotic condition

$$A_u = z_3^{1/2}A \qquad m_u^2 = m^2 - \delta_{m^2}$$

The field equation is

$$(\partial_\mu \partial^\mu + m_u^2)A_u(x) = - H_i'(A_u(x))$$

Together with the canonical commutation relation for A_u it leads[12] to

$$-iKT_x\{j\} = -z_3^{1/2} H_i'\Big(-iz_3^{1/2}\frac{\delta}{\delta j(x)}\Big)T\{j\} -$$
$$(z_3 - 1)(-iKT_x\{j\}) - iz_3\delta_{m^2}T_x\{j\} + j(x)T\{j\} \qquad (45)$$

where the second and third term on the right-hand side are the amplitude and mass renormalization terms, respectively. Equation (45) may with (25), (30), and (33) also be written

$$KR_x = - z_3^{1/2}H_i'(z_3^{1/2}R_x) - (z_3 - 1)KR_x + z_3\delta_{m^2}R_x + j(x) \qquad (46)$$

We see that $R_x\{j\}$ is the solution of same field equation as $A(x)$ with (from 40) the same initial condition $A_{\text{in}}(x)$ but modified by the source term $j(x)$, which corresponds to adding in (44) $jz_3^{-1/2}A_u$ on the right-hand side. This is [cf. (13)] the reason why, for τ and r functions, the name "Green's functions" is appropriate. The operator $A_x = iT_xT^+ = TR_yT^+$ also solves the modified field equation (46) but with the final condition $A_{\text{out}}(x)$ and generates, in its Volterra expansion 'advanced' operator products. Rewriting (45) as

$$-iKT_x\{j\} = -\tilde{H}_i'\Big(-i\frac{\delta}{\delta j(x)}\Big)T\{j\} + j(x)T\{j\} \qquad (47)$$

it is not difficult to show that its formal solution is (28) with

$$\langle T\{j\}\rangle = \frac{1}{c_v}\exp\Big[-i\int dx\tilde{H}_i\Big(-i\frac{\delta}{\delta j(x)}\Big)\Big]\exp\Big[-\frac{1}{2}j\Delta_F j\Big] \qquad (48)$$

where c_v is the formal phase factor

$$c_v = \exp\Big[-i\int dx\tilde{H}_i\Big(-i\frac{\delta}{\delta j(x)}\Big)\Big]\exp\Big[-\frac{1}{2}j\Delta Fj\Big]\Big|_{j=0}$$

which merely cancels "vacuum graphs," since the expansion of (48) in powers of $\tilde{H}_i$ produces just the usual Feynman graphs, supplied with the mass- and amplitude renormalization counter terms already.

We need, however, the perturbation expansion of r-functions. Inserting (28) with (48) into (33) gives, after some manipulations, [Exercise in repeated use of (17)] (37) with

$$\langle R_x\{j\}\rangle = -\int dx'\Delta_{\text{ret}}(x-x')\tilde{H}_i'\Big(\frac{\delta}{\delta J(x')}\Big)$$

$$\times\exp\Big[-\int\tilde{H}_i'\Big(\frac{\delta}{\delta J(u)}\Big)\frac{\delta}{\delta j(u)}du+\frac{1}{2^2 3!}\int\tilde{H}_i'''\Big(\frac{\delta}{\delta J(u)}\Big)\frac{\delta^3}{\delta j(u)^3}du-+\cdots\Big]$$

$$\times\exp\Big[\frac{1}{4}J\Delta^{(1)}J+J\Delta_{\text{ret}}j\Big]\Big|_{J=0} \tag{49}$$

where $\Delta^{(1)}=i\Delta^{(+)}-i\Delta^{(-)}$. Expanding in powers of $\tilde{H}_i'$ gives Dyson's "doubled graphs."[21] From x a Δ_{ret} line leads to the vertex x', and from the j-arguments advanced lines lead to x' or u vertices as (42) requires. In the interior of the graph, each vertex has as many legs as in a Feynman graph. However, to a vertex leads one Δ_{ret} line from a later vertex (or lead three, or five, etc. lines). The remaining lines either emanate as Δ_{ret} lines to earlier vertices, or to external y-coordinates, or connect with similar "unused" lines from other vertices as $\Delta^{(1)}$ lines. There is one latest vertex, the x' one, to which all others, the u-vertices, are (possibly multiply) connected by advanced lines, while the $\Delta^{(1)}$ lines do not imply a time relation between the vertices they connect. [The u-vertices are a "set directed upward in time." The same would be true also in classical nonlinear field theory[11] where in (49) in the first exponential only the first term and in the second exponential only the second term are to be taken, with, of course, $\tilde{H}_i'$ identical with H_i' due to $z_3=1$, $\delta_{m^2}=0$.] (The construction of doubled graphs given here differs slightly from Dyson's, cf. footnote 21 of Ref. 9.) Over all possible graphs of all orders is to be summed; each graph stands for a real invariant perturbation theoretical contribution to the r-function. However, in contrast to Feynman graphs where every line pattern occurs only once, here one line pattern admits in general of several doubled graphs to be drawn on it. For example, for the theory with $H_i(A) = gA^4$, the contributions (apart from factors and graphs containing self-contractions which are eliminated by understanding A^4 as the Wick product $A^4-6A^2\Delta_F(0)+3\Delta_F^2(0)$ and graphs containing counter terms) to $r(x, y) = \Delta'_{\text{ret}}(x - y)$ in second order are

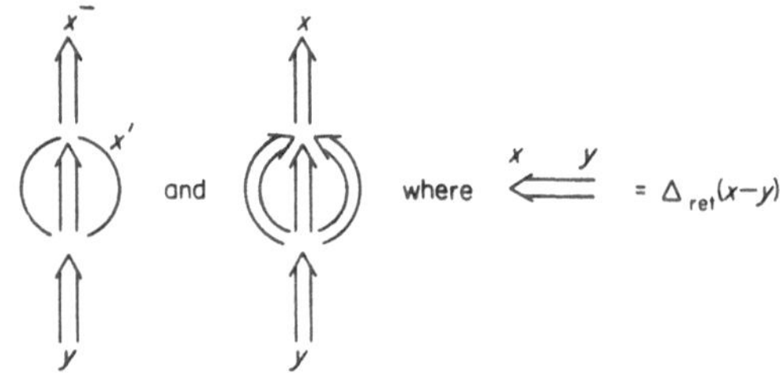

and $- = \Delta^{(1)}(x - y)$, while the third order contributions to $r(x, y_1 y_2 y_3)$ are

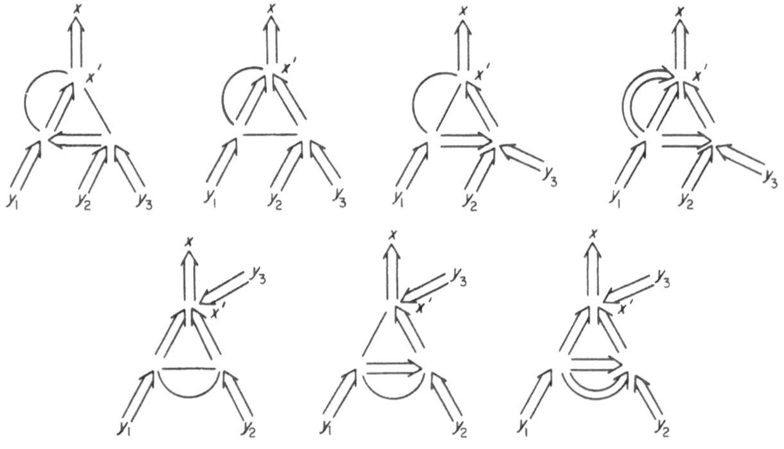

plus the graphs obtained by permutation of the y's, plus the graphs obtained by substituting the second-order corrected lines from before for one line of the first-order contribution

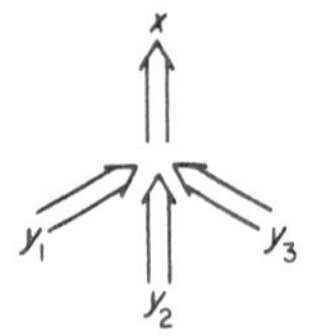

In spite of the occurrence of two types of lines, doubled graphs are, for certain purposes (namely, MPSA), more convenient than Feynman graphs. This is seen in a very simple example: To verify $T^+ T = 1$ or, explicitly,

$$\langle T^+\{j\}\rangle \exp\left[\frac{\overleftarrow{\delta}}{\delta j}\overleftarrow{K} i\Delta^{(+)}\overrightarrow{K}\frac{\overrightarrow{\delta}}{\delta j}\right]\langle T\{j\}\rangle = 1 \tag{50}$$

in some order of j, (that is, to verify the unitarity equations for τ-functions, if for these Feynman graph expansions are inserted) is simple only by transforming the integrals containing τ and $\bar{\tau}$ into a doubled graph with no latest vertex, which is, therefore, identically zero.[21,24] The reason for this simplification is that the retardedness property is easy to handle while the Feynman boundary condition that underlies the Δ_F functions is not.

(Exercise: Draw as many doubled graphs as possible for r-functions of various numbers of arguments, and meditate about what would be a sensible classification of them whereby no reference, to the order nor the actual number of legs the vertices have in the graphs, should be made. You will end up with MPSA for r-functions.)

7. AMPUTATION AND CONDENSED GRAPHICAL NOTATION

We first note that upon summing all graphs, similar as for Feynman graphs the external (and, in fact, also all internal) lines, if self energy graphs are suppressed throughout, become $\Delta'_{\rm ret}$-lines (or $\Delta^{(1)\prime}$ lines, $\Delta^{(1)\prime} = \langle\{A(x), A(y)\}\rangle$ for internal lines, which will concern us only later), which we will denote from now on as $\Longleftarrow$. It is natural to factor them out similarly as in (32):

$$r(x, y_1 \cdots y_n) = \int \cdots \int dx'\, dy_1' \cdots dy_n'\, \Delta'_{\rm ret}(x - x')$$
$$\times \Delta'_{\rm ret}(y_1' - y_1) \cdots \Delta'_{\rm ret}(y_n' - y_n) r(\underline{x_1'}, \underline{y_1} \cdots \underline{y_n'}) \tag{51}$$

Thus, underlining an argument means "amputation," which is now no longer a differentiation as in (42) but most simply defined in momentum space by dividing by the Fourier transform $\Delta'_{\rm ret}(p)$ of the $\Delta'_{\rm ret}$ function

$$\int \Delta'_{\rm ret}(x) e^{ipx}\, dx \equiv \tilde{\Delta}'_{\rm ret}(p) = \frac{1}{m^2 - p^2 - i\varepsilon p_0} + \int_{4m^2}^{\infty} \frac{\rho(\kappa^2)\, d\kappa^2}{\kappa^2 - p^2 - i\varepsilon p_0}$$
$$\varepsilon \to +0 \tag{52}$$

where $\rho(\kappa^2)$ is the Lehmann weight function.[22] [In (52) a first technical assumption is made: that the integral converges without subtraction. About dropping this technical assumption cf. Ref. 9 Section 5.]

Since $\rho \geq 0$, $\tilde{\Delta}'_{\text{ret}}(p)$ may vanish once for p^2 between m^2 and $4m^2$ and any number of times for $p^2 > 4m^2$. These are "C.D.D. zeros."[23] Although the amputated r-functions then is still well defined, since we of course require it to have the same retardedness properties as the unamputated function, it may be singular at those momentum values. (The technical assumptions that it should not be singular there seem a little drastic, so we will not make them.) "C.D.D. zeros" may occur, e.g., whenever two particles have the same quantum numbers and can convert virtually into one another. As an example, consider the Lagrangian

$$L = \frac{1}{2}\,\partial^\mu A \partial_\mu A - \frac{m_a^2}{2}\,A^2 + \frac{1}{2}\,\partial^\mu B \partial_\mu B$$

$$-\frac{m_b^2}{2}\,B^2 - \frac{g_a}{2}\,A^4 - \frac{g_b}{4}\,B^4 - g_{ab}\,AB$$

The A as well as the B-propagator may have "C.D.D. zeros," and the amputated functions then have (in perturbation theory) the corresponding singularities. However, both the A- and B-propagators will have (under some natural assumption) two one-particle poles, and forming by linear combination two new fields each associated with only one of the stable particles will in general (as is the case for $g_a = g_b = 0$) make the C.D.D. poles and singularities disappear, which have, therefore, been due only to an "unnatural" choice of fields. (Exercise: Verify the plausibility of the occurrence of "C.D.D. zeros" in this example and of their disappearance.) We shall represent graphically the amputated function as a shaded circle with a little circle for the x' coordinate and mere short strokes for the y' coordinates: Then from (34a) and (51),

$$\equiv \int \Delta'_{\text{ret}}(x-y)\,j\,(y)\,dy + \langle R'_x \rangle = \quad + \quad \bigcirc \tag{53}$$

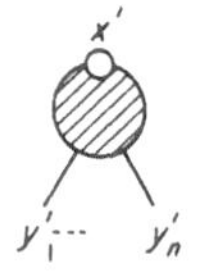

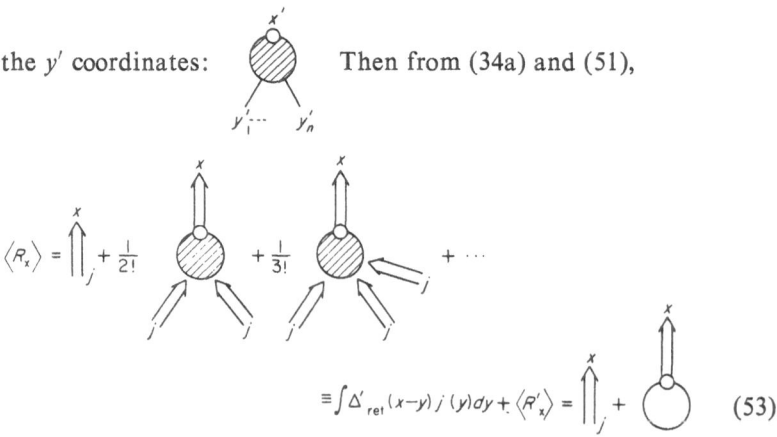

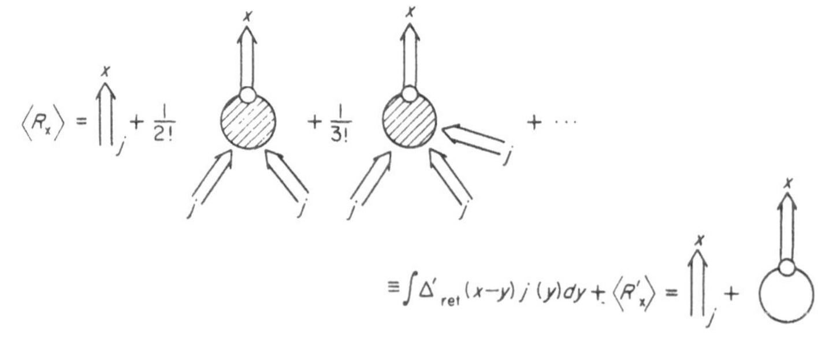

with $\langle R_{x,y_1,\ldots,y_n}\rangle =$  , where no shading means that the object is really a functional that has potentially any number of legs that can be pulled out by functional differentiation with respect to j. Note that  must have at least two y-points in order not to vanish identically, and that 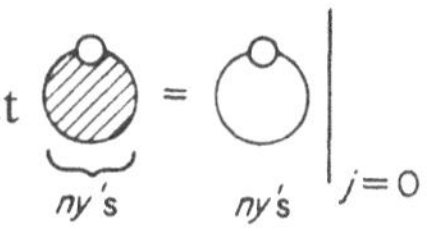

In order to translate (39b) into the present notation, we introduce  for $i\Delta^{(+)}(x-y)$. Noting that from (52)

$$\delta(p^2 - m^2)\,[(m^2 - p^2)\tilde{\Delta}'_{\mathrm{ret}}(p)] = \delta(p^2 - m^2)$$

(i.e., on the mass shell amputation is equivalent to application of the Klein-Gordon operator), using the decomposition (53) of $\langle R_x\rangle$ into the linear and the nonlinear part in j, and correspondingly separating in (39b) the j-dependent from the j-independent part, we obtain with  $= \Delta(x-y)$

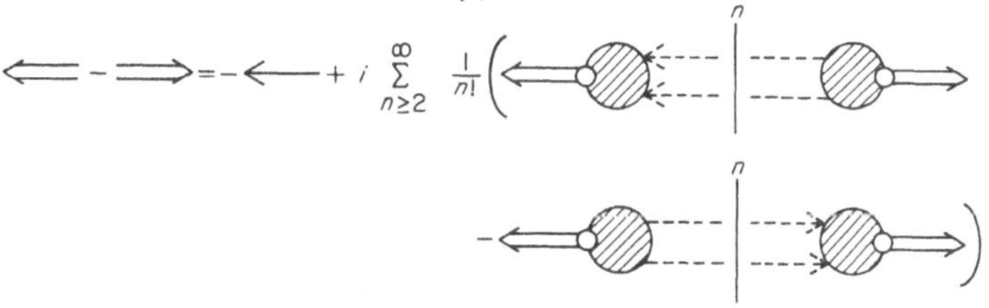

(54)

and

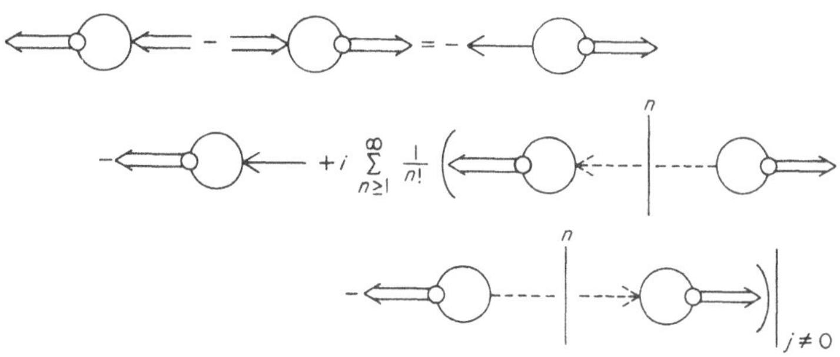

(55)

where as in (53) and in the previous section all interior points of some graphical construction are to be integrated over. Note that the last graphical equation stands for an infinite set of equations in  functions. For example, if we differentiate k times and then set $j = 0$ the bilinear expressions each decompose into 2^k terms for each n (for $n = 1$ actually only in $2^k - 2$ terms because of ). Remember also that for finite external momenta only finitely many terms actually contribute such that then no convergence problem is present here [in contrast to the situation if we had transcribed (32) into τ-functions equations].

Equation (54) is simply

$$\Delta'_{\text{ret}}(x - y) - \Delta'_{\text{av}}(x - y) = i\langle[A(x), A(y)]\rangle$$
$$= i\int dp(e^{-ip(x-y)} - e^{ip(x-y)})\langle A(0)\delta(p - \mathscr{P})A(0)\rangle \qquad (56)$$

from $A(x) = e^{i\mathscr{P}x}A(0)e^{-i\mathscr{P}x}$ where $\mathscr{P}$ is the energy-momentum operator. With

$$(2\pi)^3\langle A(0)\delta(p - \mathscr{P})A(0)\rangle \equiv \rho(p^2)$$

we obtain (52), under the technical assumptions mentioned there.

Equation (55) is, similarly, an equation for a certain absorptive part of the  function on the left-hand side.

Namely, $r(x, y, y_1 \cdots y_n)$ is the Fourier transform of an analytic function:

$$\int \exp\,[ip(x - y) + ip_1(x - y_1) + \cdots ip_n(x - y_n)]$$
$$\times\; r(x, y\, y_1 \cdots y_n) dy\, dy_1 \cdots dy_n = \tilde{r}(p\, p_1 \cdots p_n) \qquad (57)$$

is analytic in the tube: $\operatorname{Re} p$, $(\operatorname{Re} p_i)$ arbitrary, $\operatorname{Im} p$, $(\operatorname{Im} p_i)$ in the forward light cone,

$$\int \exp\,[ip(x - y) + ip_1(x - y_1) + \cdots + ip_n(x - y_n)]$$
$$\times\; r(y, x\, y_1 \cdots y_n)\, dy\, dy_1 \cdots dy_n$$
$$= \tilde{r}(-p - p_1 \cdots - p_n, p_1 p_2 \cdots p_n) \qquad (58)$$

is analytic in the tube: $\operatorname{Re} p$, $(\operatorname{Re} p_i)$ arbitrary, $\operatorname{Im}\,(-p - p_1 \cdots - p_n)(\operatorname{Im} p_i)$ in the forward light cone. For $p_1 \cdots p_n$ real, $r(p\, p_1 \cdots p_n)$

is still analytic for $\operatorname{Im} p$ in the forward light cone while $\tilde{r}[(-p - p_1 \cdots - p_n) p_1 \cdots p_n]$ is then analytic for $\operatorname{Im} p$ in the backward light cone. Equation (55) admits only all $p, (p_i)$ to be real, which is the intersection of the domains of definition of (57) and (58). The right-hand side certainly vanishes for p (real and) sufficiently spacelike if $p_1 \cdots p_n$ are fixed, since the intermediate state sum allows only timelike momenta to be transported between the two factors. To such a situation the edge of wedge theorem[25] applies. It states that the two analytic functions of p, boundary values of which stand on the left-hand side, actually continue each other analytically. Thus (57) and (58) are the same analytic function of p, and for real p two boundary values of that function. If Re p is such that the right-hand side of (55) is nonzero, we have a discontinuity of that analytic function expressed by a sum over intermediate states. This is just the generalization of the discontinuity formula (56), which after Fourier transformation reads

$$\begin{aligned} \tilde{r}(p) - \tilde{r}(-p) &= \tilde{\Delta}'_{\text{ret}}(p) - \tilde{\Delta}'_{\text{av}}(p) \\ &= 2\pi i \rho(p^2) \varepsilon(p_0) \end{aligned} \tag{59}$$

with $\tilde{r}(-p) = \overline{\tilde{r}(p)}$, or $= \overline{\tilde{r}(\bar{p})}$ for p nonreal. In connection with formulas like this, which are very similar to those one has in dispersion relations, the concept of absorptive part is familiar, in fact the derivation of dispersion relations (e.g., Lehmann[26]) makes use of formulas of the variety (39b), i.e., (55).

Actually, one may split the absorptive part on the right-hand side of (55) and introduce "interpolating" analytic functions that coincide with one another in domains larger than that described above for (58) and (57) such that all momenta but one may remain complex. They are then, by a reasoning analogous to that above, all different boundary values of one analytic function. These are the generalized retarded functions of Steinmann, Ruelle, and Araki,[27] which I will mention later once more, and which were used in Ref. 4.

The right-hand side of (55) consists of terms of singular behavior. The first two terms have one-particle δ-type end-line singularities (i.e., their Fourier transforms are proportional to $\delta(q^2 - m^2)$, q equals p, respectively, $-p - p_1 - \cdots - p_n$) besides retarded (or advanced) singularities in the other variable. The other terms give in addition to end-line singularities $\delta[(\sum' p)^2 - m^2]$ singularities from one-particle intermediate states $[(\sum' p)$ is a partial sum of the momenta in (57) or (58)], while many-particle intermediate states lead to threshold singularities. These may occur either for an end-line or for an inner momentum. Since we know the singularity structure of the end lines due to (54), for consistency the $\delta[(\sum' p)^2 - m^2]$ singularities must be reproduced in (55) on the left-hand side. This we shall prove now, and moreover, that also all one-particle singularities of δ- and retarded and advanced types in (55) cancel. While of course, this must be so, and furnishes us only information we can obtain also without the nonlinear system, the corresponding information concerning two- (or more) particle singularities is rather less trivial and will, in fact, require a specially adapted apparatus to extract.

8. ONE-PARTICLE STRUCTURE

From your occupation with the exercise given at the end of Section 6, you know that (amputated) doubled graphs, like Feynman graphs, can be classified into those that can be separated into two parts by cutting one (necessarily doubled) line, and those that cannot. Those that can be so separated we call one-particle reducible, the others one-particle irreducible. The reducible ones can be cut, so as to decompose, in general in several ways. We classify these graphs further into those that admit one-particle cuts between the latest argument and a fixed number, $n \geq 1$, of earlier arguments, these n taken as a whole, and those that don't. Calling the latter ones one-particle irreducible with respect to the described configuration, indicated by a broken line between the two groups of coordinates, and noting that in the reducible graphs we have to avoid double-counting, we are led to the formula

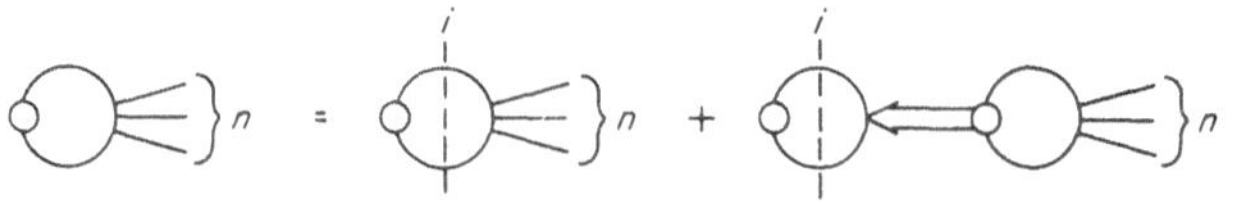

(60)

which in fact defines 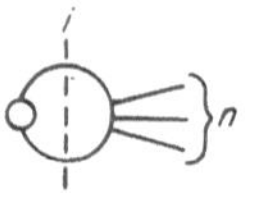 for $n > 1$ if it is known for $n = 1$.

Equation (60) for $n = 1$ allows the iteration solution

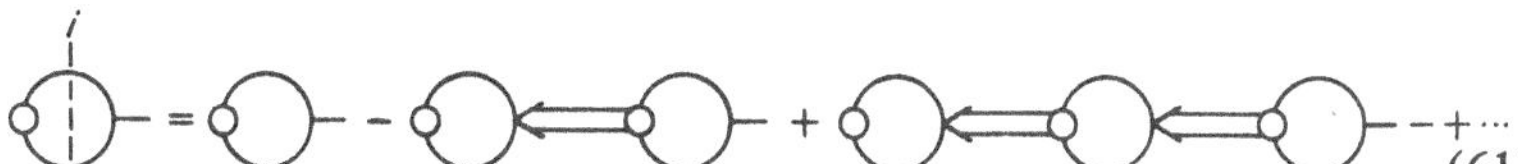

(61)

since due to 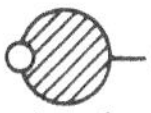 $\equiv 0$ for any finite number of arguments (or, to any finite j-order) the right-hand side consists of finitely many terms only. Hereby equation (61) can be resummed to

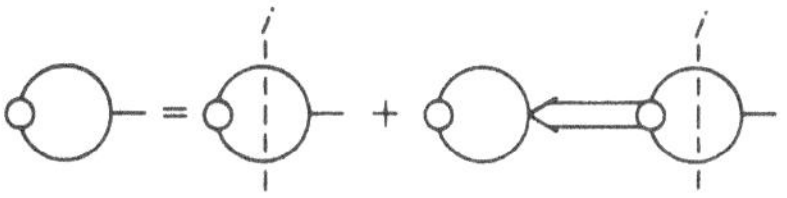

(62)

"Multiplying" (60) from the left by 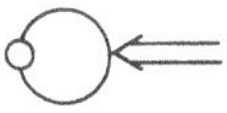 we obtain with (62)

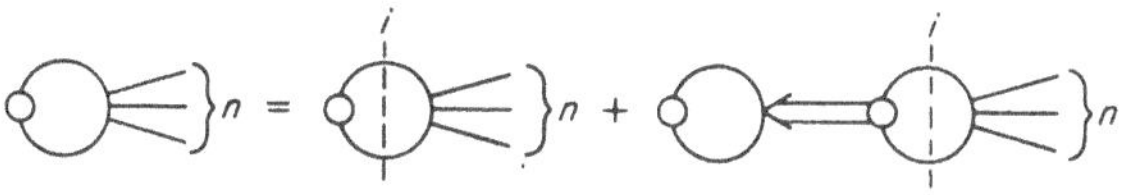

(63)

Note that, while (60) displays the first one-particle reducibility (singularity in momentum space) from the left, (63) does so from the right.

We have not paid attention here to the C.D.D.-singularities that the amputated functions may possess. Actually, they do not make (60) ambiguous, since

$$\frac{1}{x + i\varepsilon}\, x\, \frac{1}{x + i\varepsilon}\bigg|_{\varepsilon\to+0} = \frac{1}{x + i\varepsilon}\bigg|_{\varepsilon\to 0} \tag{64}$$

and this equation also makes the manipulations that lead to (63) correct. We must, however, keep in mind for later that one-particle irreducible functions may possess "inner" C.D.D.-singularities. In fact, from (60) and (61) we have

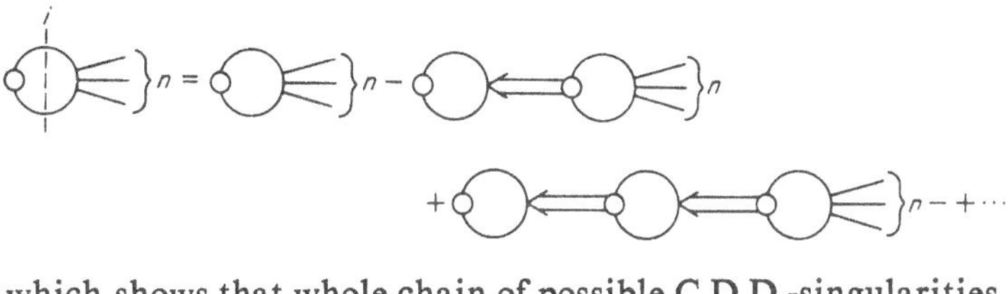

(65)

which shows that whole chain of possible C.D.D.-singularities. Note also that because of 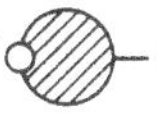 $\equiv 0$ the momenta transferred along the one-particle bridges in (65) will (at least if never sharp external momenta are used) all be different from one another.

We shall now show that in (55) all one-particle reducibilities (which may or may not be actual singularities for real p) between the two displayed arguments cancel. [If the momentum transported between those two arguments is p, these will be in Fourier space, e.g., singularities proportional to $\delta[(p+\cdots)^2 - m^2]$ or to $[m^2 - (p+\cdots+i\varepsilon)^2]^{-1}$ or to $[m^2 - (p+\cdots-i\varepsilon)^2]^{-1}$. Here, ε is an infinitesimal positive timelike vector, and ... stands for a partial sum of the other momenta that are suppressed in the notation.] Namely, (60) and (62) or (63) give

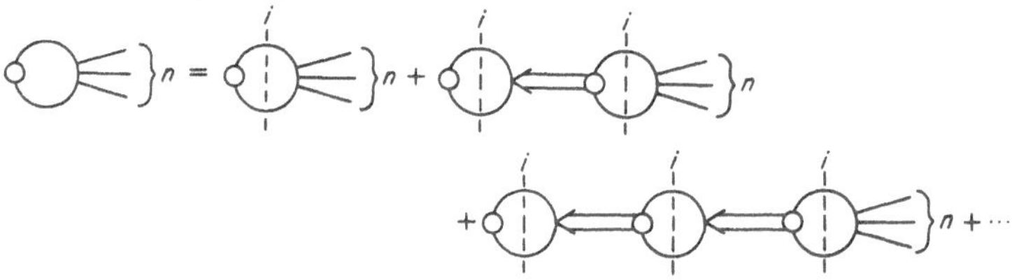

(66)

Nonsingular terms are those where momentum conservation makes the respective function zero at the point where otherwise a singularity would be expected, e.g., in (54) all terms in the sum over n are nonsingular, namely, zero, at $p^2 = m^2$. (In two-particle structure analysis, reducibilities may not lead to singularities also for other reasons.) We insert equations (60) and (63) into (55) in such a manner as to display the first one-particle reducibility from the left:

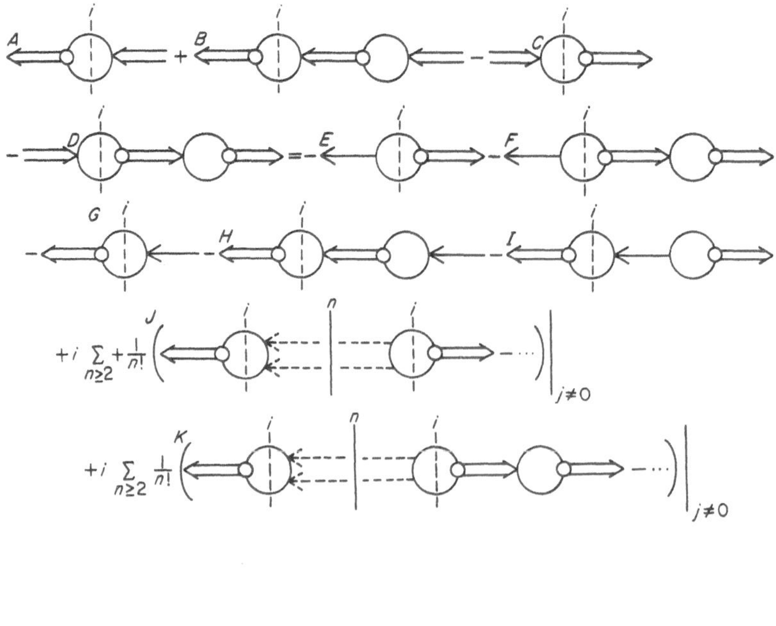

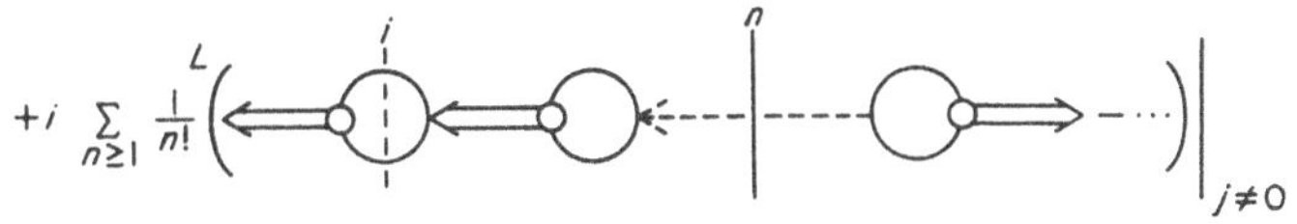

(67)

Note that in K and L the restriction $j \neq 0$ may be dropped since, due to 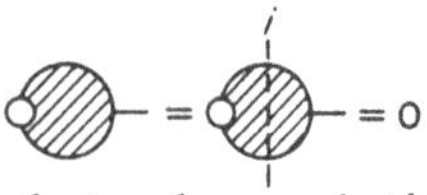 if $j = 0$, K and L are zero anyway. This means that when substituting (55) in L we have to take (54) also:

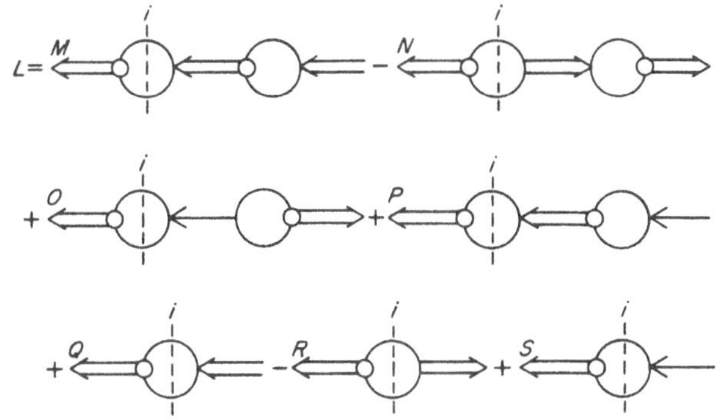

(68)

Cancelling M/B, O/I, P/H, Q/A, S/G and separating K with the help of (54),

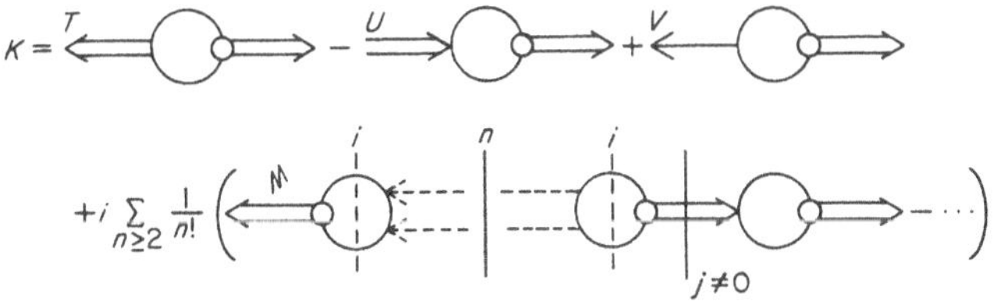

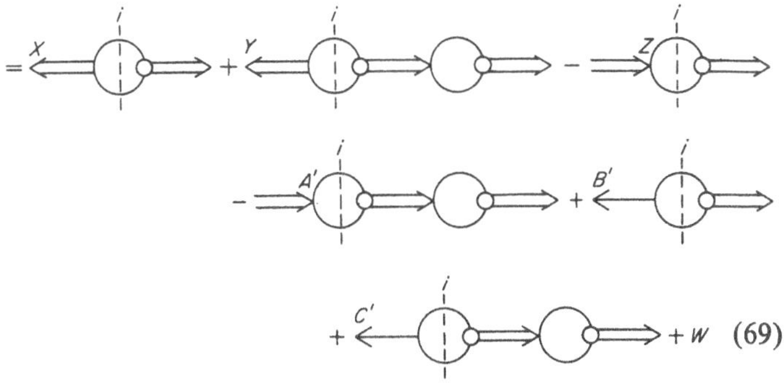

(69)

and cancelling Z/C, A'/D, B'/E and C'/F we can collect the remaining terms J, X, Y, W, N, and R into

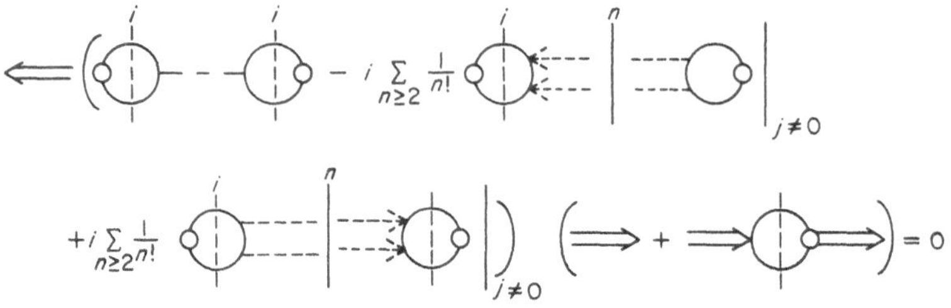

(70)

Multiplying from the right by 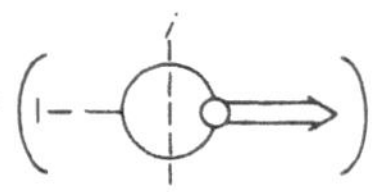 and using (60) for $n = 1$ eliminates in (70) the last term in the last bracket. Amputation then leads to the result that the first bracket in (70) vanishes. This bracket closely resembles (55), with, however, substitution of one-particle irreducible functions, omission of one-particle intermediate states, and neglect of endlines. [That the "end line structure" must cancel out for consistency was already remarked in Section 7 and is obvious in the stronger result (70).] Thus, we may say for brevity that the absorptive part of one-particle irreducible functions is a bilinear construction analogous to that for the original functions, with one-particle intermediate states omitted. The mass threshold has been raised to two-particle states. We may anticipate, and this will be verified later, that also two-particle intermediate states can be eliminated similarly and the mass threshold raised to three-particle states. Mere technical complications made it uninviting to go far beyond that, but with complete confidence we may say that it is possible to raise the mass thresholds to an arbitrarily high state. The picture that evolves is MPSA of Green's functions.

Coming back to (70), we note that amputation is an unambiguous operation only if either $\tilde{\Delta}'_{\text{ret}}(p)$ does not vanish for real p (i.e., has no C.D.D. zeros) or if the retarded boundary condition is available as remarked in connection with equations (51) and (64), (65). It is not available in (70). Moreover, if $\tilde{\Delta}'_{\text{ret}}(p)$ does have C.D.D. zeros, e.g., the term N in (68) is ambiguous since in contrast to (64)

$$\left[\frac{1}{x+i\varepsilon}\,x\right]\frac{1}{x-i\varepsilon}\bigg|_{\varepsilon\to+0} \neq \frac{1}{x+i\varepsilon}\left[x\,\frac{1}{x-i\varepsilon}\right]\bigg|_{\varepsilon\to+0},$$

and, in fact, (65) leads us to suspect that the first bracket in (70) can then not vanish since "inner" C.D.D.-singularities must give rise to additional discontinuities of the analytic functions apart from those written in (70). I shall give only the final result and suppress its slightly tedious proof. Let $\overset{-1}{\Longleftarrow}$ be the inverse of Δ'_{ret} under observation of retardedness, i.e., the kernel with which amputations are to be carried out. Then (54) becomes

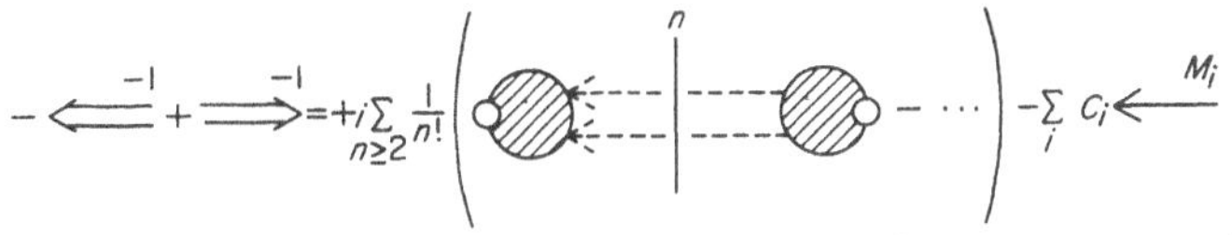

(71)

where $\sum_i$ goes over the C.D.D. zeros at $p^2 = M_i^2$, the c_i are positive, and $\underset{x \quad y}{\overset{M_i}{\longleftarrow}}$ is $\Delta_{M_i}(x-y)$. Equation (70) is to be replaced by

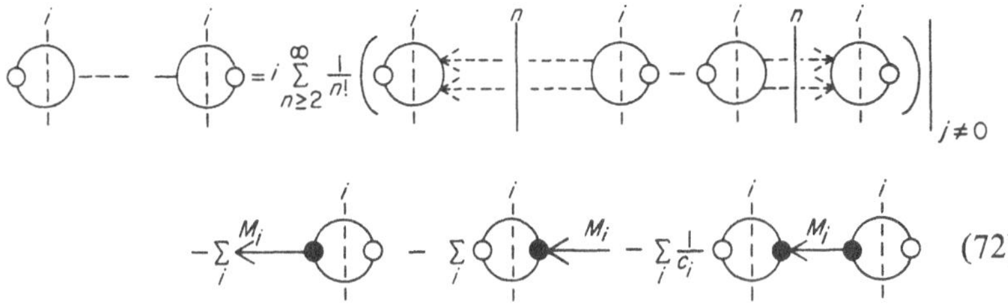

(72)

where a fattened connection means that the Klein–Gordon operator $(\Box + \cdots M_i^2)$ is applied to that argument of the irreducible amputated function, whereby it becomes nonsingular at $p^2 = M_i^2$ such that Δ_{M_i} can be joined there. [The last term on the right-hand side of (72), i.e., the effect of inner C.D.D.-singularities, was overlooked in the statement in Ref. 11 after (IV. 12) about the form of $\tilde{B}_{pq}$. [The technical assumption that amputated functions are not C.D.D.-singular leads, of course, to (IV. 13) of Ref. 11.]

It remains to show that the 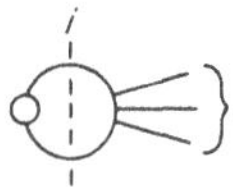 functions do not have any one-particle singularity in the momentum $(p + \cdots)$, where p is the momentum transported between the left point and the right points, and $+ \cdots$ stands for a partial sum of the suppressed momenta. Clearly, it suffices to discuss $n = 1$. This can be done in several ways: a) One may iterate (72) and then find that if there would be a singularity at $(p + \cdots)^2 = m^2$, it could occur only where at least two such relations with different partial sums are satisfied. A corresponding singularity is, therefore, of an entirely different character as the original one-particle singularities, which are, therefore, correctly displayed in (66). b) One may show on the axiomatic basis alone that in (65) all one-particle singularities [at $(p + \cdots)^2 = m^2$] cancel on the right-hand side. This proof is essentially that given at the beginning of Section 2 of Ref. 9 and depends on the following generalization of (21),

$$R[x, (x + \xi_1) \cdots (x + \xi_n)] - \int \Delta_{\text{av}}(x - x')\vec{K}R[x', (x' + \xi_1) \cdots (x' + \xi_n)]\, dx' = \int A_{\text{out}}(x - \xi)F(\xi, \xi_1 \cdots \xi_n)\, d\xi$$

where $F(\xi, \xi_1, \cdots \xi_n)$ is a C-number. This is a special case of a much more general formula due to Haag.[28] c) One may regard this statement as a specialization of a more general statement to be discussed next.

We replace (66) by an ansatz that is to introduce the "entirely one-particle irreducible" functional :

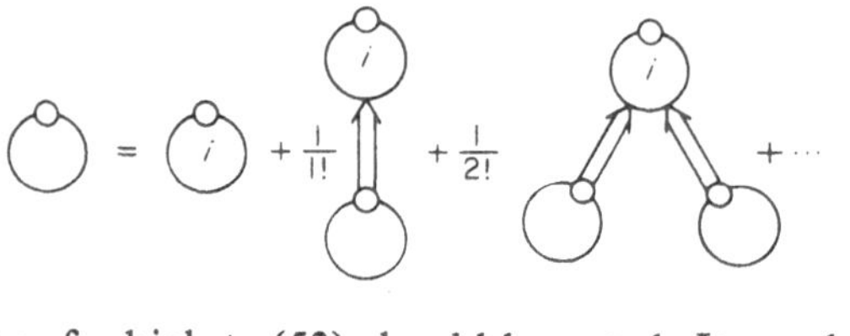

(73)

the similarity of which to (53) should be noted. It can be solved for  by iterating

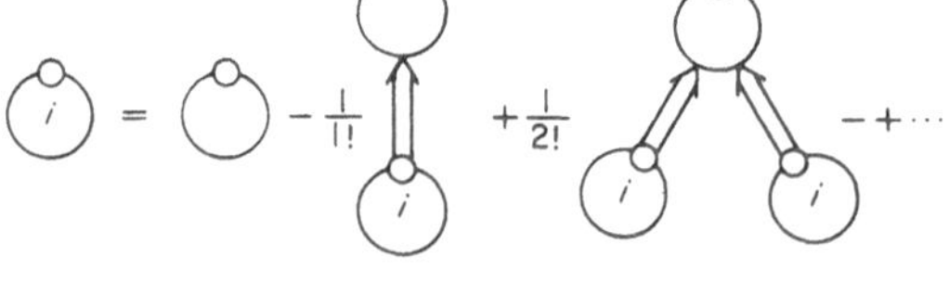

(74)

which follows, Ref. 9, from (73). The resulting expression, which should be compared with (65), has for any finite order n (i.e., only a finite number of arguments is suppressed) only a finite number of terms due to 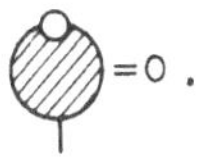. Using the previous result for  or the known singularity structure for , one may show that  has no one-particle singularity (which necessarily would have to be a retarded one since  is a retarded functional) with respect to any separation of arguments into two groups. On the other hand, one also may prove, directly by induction on the number of arguments, that the irreducible functions are free of one-particle singularities even if the generalized retarded functions mentioned earlier are included.[6] [Streater's paper consists of two parts. The first concerns the proof just mentioned and uses the fact that for generalized retarded functions the discontinuities on the right-hand-side of (55) are separated entirely, i.e., the coincidences mentioned before are avoided; that part is marred by misprints only.]

In Ref. 6 and in Ref. 29, where Zimmermann performs a similar analysis for τ-functions, amputation is done by applying the Klein-Gordon operator, and connecting links are correspondingly Δ_{ret} instead of Δ'_{ret} functions. Our method can be easily adapted to this case also, where, in fact, the C.D.D. complication is avoided. However, this method does not seem to have a natural generalization to the several-particle case.

For completeness I will mention that in Ref. 9 also the functionals

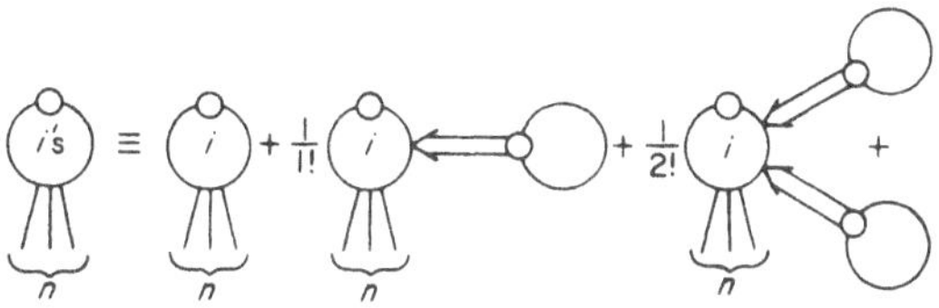

are dealt with; they differ from  by being one-particle irreducible for any division of the $n + 1$ displayed coordinates into two groups instead of for only one particular division.

9. TWO-PARTICLE STRUCTURE

Inspection of the doubled graphs described in Section 6 shows that two-line cuts between any two parts of a doubled graph, such that the latest coordinate is to the left, can be of three types

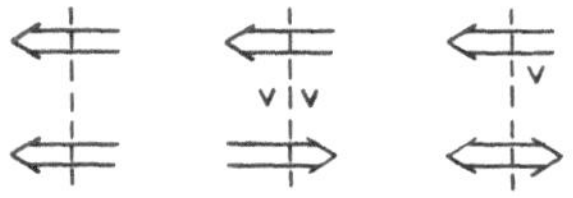

(75)

where the sign $\vee$ means that (necessarily) the lower argument on that side of the cut will be timelike earlier than the upper argument. This follows simply because any nonvanishing graph part must have a latest vertex (which in our condensed notation means that it must have a -argument), and because amputation as explained earlier does not change retardedness properties. The lower line in the third link is the "corrected contraction line" $\Delta^{(1)'}$ mentioned at the beginning of Section 7.

The first cut listed in (75) has on the right a graph part with two latest vertices, from which two "trees" of retarded lines will start. However, these trees also may be connected with each other by contraction lines $\Delta^{(1)}$ or even retarded lines, since according to (49) to one vertex any odd number of retarded lines may lead, not only one. Such graphs or functions are as yet undefined; inspection of (49) shows that we should expect them to be the generalized retarded functions

$$r(x_1\cdots x_m, y_1\cdots y_n) = \langle R(x_1\cdots x_m, y_1\cdots y_n)\rangle \tag{76a}$$

where

$$R(x_1\cdots x_m, y_1\cdots y_n) = R_{x_1\cdots x_m, y_1\cdots y_n}\{0, 0\} \tag{76b}$$

and

$$R_{x_1\cdots x_m, y_1\cdots y_n}\{J, j\} = \prod_{k=1}^{m} \frac{\delta}{\delta J(x_k)} \prod_{i=1}^{n} \frac{\delta}{\delta j(y_i)} R\{J, j\} \tag{76c}$$

with

$$R\{J, j\} = \left(T\left\{j - i\frac{J}{2}\right\}\right)^{+} T\left\{j - i\frac{J}{2}\right\} \tag{77}$$

which, e.g., by the familiar Feynman's disentangling method, is easily seen to be

$$R\{J, j\} = \bar{T} \exp\left[\frac{1}{2}\int J(x)R_x\{j\}\,dx\right] T \exp\left[\frac{1}{2}\int J(x)R_x\{j\}\,dx\right] \quad (78)$$

where T and $\bar{T}$, mean time ordering, respectively, anti-time ordering with respect to the x-argument only, i.e., the functionals are treated as a whole (cf. also Ref. 15). From (78) follows that $R(x_1 \cdots x_m, y_1 \cdots y_n) = 0$ unless each y is timelike earlier to at least one x.

Like τ-functions, generalized r-functions will, in general, decompose into a connected part and several disconnected ones. The latter ones will be characterized in momentum space by the presence of factors $\delta(\sum' p_{x_i} + \sum'' q_{y_i})$, $\sum'$ and $\sum''$ being partial sums, besides the overall δ-function $\delta(\sum_{i=1}^{m} p_{x_i} + \sum_{j=1}^{n} q_{y_j})$, with $\sum'$ nonempty. Just as for connected τ-functions, the generating functional is

$$\ln \langle T\{j\}\rangle \equiv \langle T\{j\}\rangle_C$$

[as learned, for instance, from the vacuum expectation value of (45) or, more generally, Ref. 28], for connected r-functions it is

$$\ln \langle R\{J, j\}\rangle \equiv \langle R\{J, J\}\rangle_C \quad (79)$$

the subscript C meaning "connected," as seen for instance, from the analog of (45) for R-functionals given in Ref. 9. (Actually, for any product of T- and T^+-functionals we have

$$\ln \langle T\{\cdots\}T^+\{\cdots\}T\{\cdots\}T\{\cdots\}\cdots\rangle$$
$$= \langle T\{\cdots\}T^+\{\cdots\}T\{\cdots\}T\{\cdots\}\cdots\rangle_C$$

This formula is related to that for cumulants in many-body problems.)

Inserting (28) or (37) into (77), respectively, (78) gives

$$R\{J, j\} = :e^{\Delta_{\text{in}}\overleftrightarrow{K}(\delta/\delta j)}: \langle R\{J, j\}\rangle \quad (80)$$

and this gives with (28) and (48) for the perturbation theoretical expansion

$$\langle R\{J, j\}\rangle = \exp\left[-\int \tilde{H}'\left(\frac{\delta}{\delta J(u)}\right)\frac{\delta}{\delta j(u)}\,dv + \cdots\right]$$
$$\times \exp\left[\frac{1}{4} J\Delta^{(1)}J + J\Delta_{\text{ret}}j\right] \quad (81)$$

where the exponents are the same as in (49). Equation (78) shows that

$$R_x\{j\} = R_x\{J, j\}|_{J=0}$$

where $R_x\{j\}$ is the same as in (33) or (34a), and

$$\begin{aligned} R_{x_1\cdots x_m}\{j\} &= R_{x_1\cdots x_m}\{J, j\}|_{J=0} \\ &\equiv R_{x_1\cdots x_m} \end{aligned}$$

correspondly generates generalized retarded products.

The functions (76) are not related to the generalized retarded functions (Ref. 27) mentioned in Sections 7 and 8. They are, except for $r(x, y_1 \cdots y_n)$, not boundary values of analytic functions. However, their introduction is unavoidable for MPSA. The identification of what stands on the right-hand-side of the first cut in (75) as such a generalised retarded function is correct, and is shown by the success of two-particle structure analysis based on the corresponding ansatz.

From generalized r-functions by simple linear combination τ or $\bar{\tau}$ functions, or vacuum expectation values of certain other operator products, are obtained. Indeed, from (77) we have

$$\begin{aligned} &(-i)^{l-k}\Big\langle \Big(T_{u_1\cdots u_k}\Big\{j - i\frac{J}{2}\Big\}\Big)^{+} T_{v_1\cdots v_l}\Big\{j - i\frac{J}{2}\Big\}\Big\rangle \\ &= \prod_{i=1}^{k}\Big(\frac{\delta}{\delta J(u_i)} + \frac{i}{2}\frac{\delta}{\delta j(u_i)}\Big)\prod_{j=1}^{l}\Big(\frac{\delta}{\delta J(v_j)} - \frac{i}{2}\frac{\delta}{\delta j(v_j)}\Big)\langle R\{J, j\}\rangle \end{aligned} \tag{82}$$

The result is that this relation permits us to transcribe integral relations between generalized r-functions into similar ones for τ, $\bar{\tau}$ and mixed functions $\langle \bar{T}A(v_1)\cdots A(u_k)\cdot TA(v_1)\cdots A(v_e)\rangle$ and that in the simplest case, if one is interested in τ-functions only, all other functions drop out. This I shall show later.

We now introduce an appropriate graphical notation for these generalized retarded functions. We shall, of course, use bubbles for connected functions, but the question of how to extract end lines from doubled graphs is in the generalized case not as trivial as in Section 7. If we have, e.g., two x-coordinates in (76), then the external lines (with self-energy parts already summed up), as inspection of (81) shows, will be

$$\begin{matrix}\Longleftarrow \\ \Longleftarrow\end{matrix} \quad \text{or} \quad \begin{matrix}\Longleftarrow \\ \Longleftrightarrow\end{matrix} \quad \text{or} \quad \begin{matrix}\Longleftrightarrow \\ \Longleftarrow\end{matrix} \quad \text{but not} \quad \begin{matrix}\Longleftrightarrow \\ \Longleftrightarrow\end{matrix}$$

since what stands to the right would not have a latest vertex in

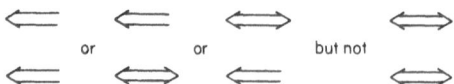

the last case. However, we may set

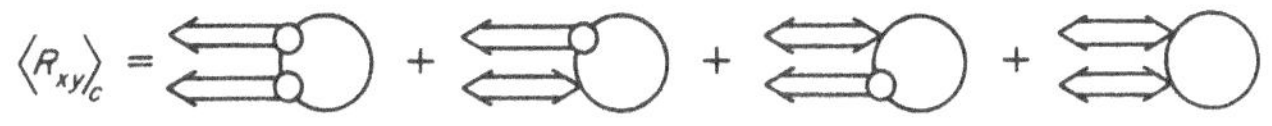 (83)

since the last term is identically zero due to $J = 0$. (83) defines what we shall mean by  .

It turns out to be convenient not just to amputate Δ'_{ret} but the whole one-particle reducible external part of the graph, i.e., the structure 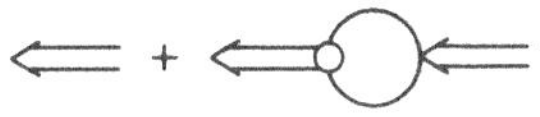 the inverse of which is 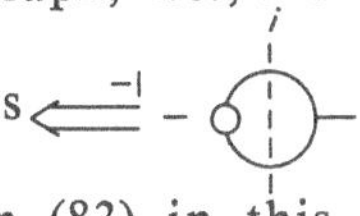 due to (60) and (62). We simply interpret all lines in (83) in this new manner, and likewise $\underset{x\quad y}{\Longleftrightarrow}$ as $\langle R_{xy}\{j\}\rangle_C$ instead of $1/2\ \langle\{A(x), A(y)\}\rangle$. This is, namely, from, e.g., (78)

$$R_{xy} = \tfrac{1}{2}\{R_x, R_y\}$$

and with (79) [using now capital subscripts for legibility]

$$\langle R_{XY}\rangle_C = \tfrac{1}{2}\langle\{R_X, R_Y\}\rangle - \langle R_X\rangle\langle R_Y\rangle$$

such that the new interpretation for $\Longleftrightarrow$ is analogous to that for $\Longleftarrow$. We now proceed to write the analog of (60) for the two-particle case,

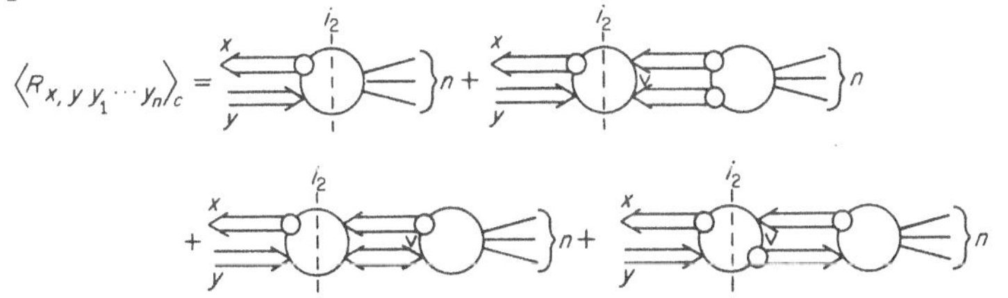 (84)

Here the V sign has the same meaning as explained after (75), except that in the second term on the right-hand side it merely splits the integration region noninvariantly; otherwise, a factor $\frac{1}{2}$ (related to Bose statistics) would have been necessary. The factors to the right of the irreducible functional in the second and third term together are, according to (83), just the (from its left) unamputated functional. Underlining the group $y_1 \cdots y_n$ means also that all common one-particle reducibilities like 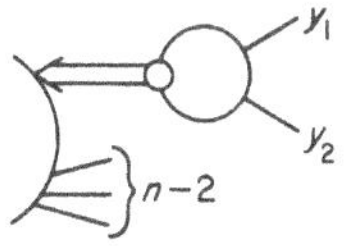 have been removed throughout. Furthermore, we will understand that in (84) all functionals are already one-particle irreducible between x, y and $y_1 \cdots y_n$ (although the case of generalized retarded functions was not dealt with in Section 8, it will not be elaborated here since it is quite straightforward).

Equation (84) contains two irreducible functionals, therefore, we will have to use another equation in addition to determine them both. For this, we need consider only $n = 2$. Clearly, the whole complex of formulas will be tedious to write or handle, so we introduce the appropriate matrix notation.

$F =$ 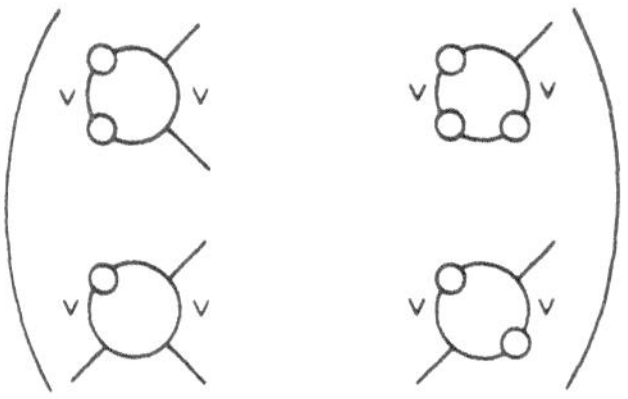

$E =$ 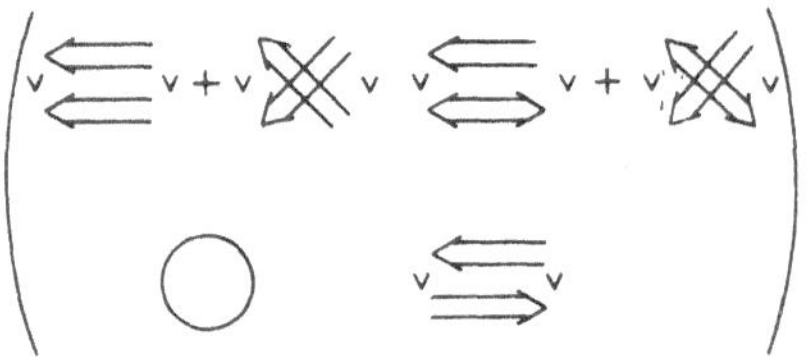

Then

$$\begin{pmatrix} \langle R_{XY,\, ZU} \rangle'_C & \langle R_{XYU,\, Z} \rangle'_C \\ \langle R_{X,\, YZU} \rangle' & \langle R_{XU,\, ZY} \rangle'_C \end{pmatrix} = G - E = EFE \qquad (85)$$

where the prime means: "one-partial irreducible" between X, Y and Z, U. The arguments have been named according to 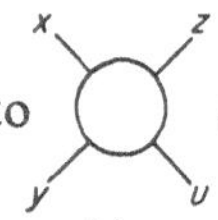 and always $X^0 > Y^0$, $Z^0 > U^0$, in general, noninvariantly. [Actually, here also $X^0 > Z^0$. G is, according to (85), the matrix with primed and connected functionals plus those disconnected ones where none of the four coordinates is disconnected from the others.] The splitting of integration regions seems unnatural and, in fact, could be avoided; however, a simple mechanization of two-particle structure analysis, which is mandatory since all others are rather more complicated, requires that splitting as we shall see. [Exercise: show that in equation (85) all interior integrations are actually restored to their complete invariant range in conformity with (83).]

We now replace (84), for $n = 2$ by the set of four equations

$$F = F_\iota + F_\iota E F \tag{86}$$

the lower row of which gives the two equations necessary to determine the irreducible functionals of (84). The upper row has to be taken along also, since we need it to be able to obtain the analog of equation (62). The solvability of (84) need be discussed only for $j = 0$, since from $F_\iota\{0\}$ we can obtain its power series expansion by iterating

$$\frac{\delta F_\iota}{\delta j} = (1 - F_\iota E) \frac{\delta F}{\delta j} (1 - E F_\iota) - F_\iota \frac{\delta E}{\delta j} F_\iota \tag{87}$$

For $j = 0$, equation (86) becomes a set of ordinary coupled integral equations of convolution type in the difference $X - Z$, with $X - Y$ and $Z - U$ as "inner coordinates" of the functions involved, and in $X - Z$ we have retardedness. This makes plausible the unique solvability of (86) for $j = 0$ and, with (87), for all j-powers if only the integration over the "inner coordinates" are so behaved as to make, e. g., a suitable generalization of the Fredholm method applicable since zeroes of the Fredholm denominator are to be treated according to the retardedness prescription. (For a discussion of this see Ref. 9.)

Actually, the Fredholm method need not work even if an integral equation has a unique solution, namely, if e.g, all traces of iterated kernels diverge. Recently, this point has been the subject of investigations in connection with extraction of Regge-poles from ladder approximations to Bethe–Salpeter equations. We will not go into this, but proceed on the technical assumption that (86) can be solved uniquely for $j = 0$. In the scalar theory with A^4-coupling in four dimensions (in contrast to fewer ones) this is, in fact, not true in renormalized perturbation theory due to logarithmic divergence of the coupling constant renormalization factor. This may be a feature of perturbation theory only (especially if $g_{\text{unren}} = 0$ would already remove the difficulty), otherwise a modification of the integral equations by subtraction procedures in the spirit of Ferretti's closed form (Ref. 30). of Salam's renormalization prescription (Ref. 31) would have to be introduced. It is this technicality that we suppress by our assumption.

With (85) we may write (86) also

$$G = E + E F_\iota G = E + G F_\iota E \tag{88}$$

since the assumed unique solvability of (86) gives also

$$F = F_\iota + FEF_\iota \tag{89a}$$

from

$$F_\iota = E^{-1} - G^{-1} \tag{89b}$$

equations that are the analog for the present situation of (61) and (62). To write them compactly, we introduce in the 2×2 matrix space of (85) the matrix $\sigma_1 = \left(\begin{smallmatrix} 0 & 1 \\ 1 & 0 \end{smallmatrix}\right)$ and denote transposition (of as well the matrix as also of the sides of the matrix elements) by superscript T. Then we have

$$\sigma_1 F^T \sigma_1 = \tilde{F} = \tag{90}$$

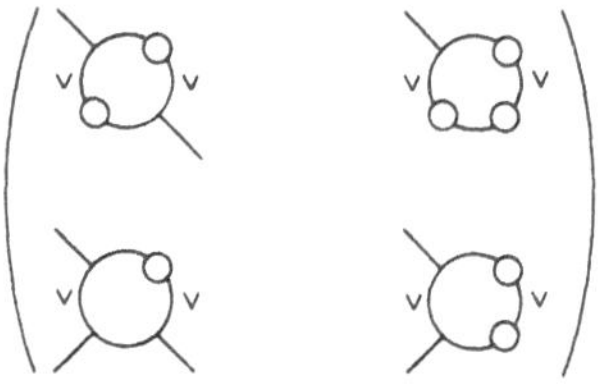

Clearly, from (88)

$$\tilde{G} = \tilde{E} + \tilde{E}\tilde{F}_\iota\tilde{G} = \tilde{E} + \tilde{G}\tilde{F}_\iota\tilde{E} \tag{91}$$

We consider

$$G - \tilde{G} - E + \tilde{E} = (E - \tilde{E})\,\tilde{F}_\iota\tilde{G} + E\,(F_\iota - \tilde{F}_\iota)\,\tilde{G} + EF_\iota\,(G - \tilde{G}) \tag{92}$$

for whose four matrix elements a straightforward calculation on the basis of (77) gives

$$\begin{aligned}(G - \tilde{G} - E + \tilde{E})'_{11} &= \langle R_{XY,ZU}\rangle'_C - \langle R_{YZ,\,XU}\rangle'_C \\ &= i\,\langle[R_{XY,U}, R_Z]\rangle'_C + i\langle[R_{XY}, R_{Z,U}]\rangle'_C \\ &\quad + \frac{1}{2}\langle\{R_{X,U}, R_{Z,Y}\}\rangle + \frac{1}{2}\langle\{R_X, R_{Z,YU}\}\rangle'_C\end{aligned} \tag{93a}$$

$$\begin{aligned}(G - \tilde{G} - E - \tilde{E})'_{12} &= \langle R_{XYU,Z}\rangle'_C - \langle R_{YZU,X}\rangle'_C \\ &= -\frac{1}{2}\langle[R_{X,YU}, R_Z]\rangle'_C + i\,\langle[R_{XY}, R_{ZU}]\rangle'_C \\ &\quad + \frac{i}{4}\langle[R_{X,U}, R_{Z,Y}]\rangle + \frac{1}{2}\langle\{R_X, R_{ZU,Y}\}\rangle'_C\end{aligned} \tag{93b}$$

$$\begin{aligned}(G - \tilde{G} - E + \tilde{E})'_{21} &= \langle R_{X.YZU} \rangle' - \langle R_{Z,XYU} \rangle' \\ &= i \langle [R_{X,YU}, R_Z] \rangle' + i \langle [R_{X.Y}, R_{Z,U}] \rangle' \\ &\quad + i \langle [R_{X.U}, R_{Z,Y}] \rangle + i \langle [R_X, R_{Z,YU}] \rangle'\end{aligned} \tag{93c}$$

and

$$\begin{aligned}(G - \tilde{G} - E + \tilde{E})'_{22} &= \langle R_{XU,YZ} \rangle'_C - \langle R_{ZU}, R_{XY} \rangle' \\ &= -\frac{1}{2} \langle \{R_{X.YU}, R_Z\} \rangle'_C + i \langle [R_{XY}, R_{ZU}] \rangle'_C \\ &\quad - \frac{1}{2} \langle \{R_{X,U}, R_{Z.Y}\} \rangle_C + i \langle [R_X, R_{ZU,Y}] \rangle'_C\end{aligned} \tag{93d}$$

These are valid, except for the familiar (93c), only for $X^0 > Y^0$, $Z^0 > U^0$. But this last restriction has put (93) into a form which invites matrix notation throughout. Nevertheless, the amputation from the left, subsequent display of the first two-particle reducibility from the left, and comparison with (92) turns out to lead to laborious calculations (though the result is not too complicated). It is: denote the difference of the two sides by X. Rewrite the same difference, but amputated from both sides and made two particle irreducible, by X^i_{amp}. Then

$$X^i_{\text{amp}} G + \eta^i_{\text{amp}} \tilde{H} = 0 \tag{94}$$

Here $\tilde{H}$ is again a 2×2 matrix, and η^i_{amp} is related to a matrix η like X^i_{amp} is to X, where the four matrix elements are

$$\eta_{11} = \langle R_{ZY,XU} \rangle'_i - \frac{1}{2} \langle \{R_{Z,UX}, R_Y\} \rangle'_C - \frac{1}{2} \langle \{R_{Z,X}, R_{Y,U}\} \rangle \tag{95a}$$

$$\eta_{12} = \langle R_{ZU,YX} \rangle'_C - \frac{1}{2} \langle \{R_{ZU,X}\ R_Y\} \rangle'_C + \frac{i}{4} \langle [R_{Z,X}, R_{Y,U}] \rangle \tag{95b}$$

$$\eta_{21} = \langle R_{Z,UXY} \rangle'_C - i \langle [R_{Z,UX}, R_Y] \rangle' - i \langle [R_{Z,X}, R_{Y,U}] \rangle \tag{95c}$$

$$\eta_{22} = \langle R_{ZU,XY} \rangle'_C - i \langle [R_{Z,U,X}, R_Y] \rangle'_C - \frac{1}{2} \langle \{R_{ZX}, R_{Y,U}\} \rangle_C \tag{95d}$$

of which the first terms are $\tilde{F}$. We have $\eta = 0$ identically as before $X = 0$, but $\eta^i_{\text{amp}} = 0$ need be shown separately. The appropriate analysis gives

$$\eta^i_{\text{amp}} \tilde{G} = 0 \tag{96}$$

Assuming, as before [in (89b)] that $\tilde{G}$ has an inverse (96) gives η^i_{amp}

$= 0$ and then (94) $X^{t}_{\text{amp}} = 0$, which means that the nonlinear systems do hold in the expected "reduced" form, analogous to (72), for irreducible functions.

Actually, a number of technical assumptions went into this proof, since, otherwise, we also should have obtained correction terms as in (72). Even if F_t from (89b) exists as assumed, it may be singular and, therefore, the associativity of the products in (92) in general violated as in the relation before (71) and then also from (96), one may only conclude that $\eta^{t}_{\text{amp}} = 0$ unless the momentum between x, y and z, u [as shown after (87)] is such that G^{-1} is singular. Taking in (96) and (94) these possibilities into account would restore the similarity to (72). However, the precise form of these "two-particle C.D.D.-singularities" and their contributions have not been analyzed yet.

How does one separate out the first two-particle reducibility from the left in brackets such as appear in (93) and (95)? For example, for brackets such as $\langle [R_{x,y}, R_{z,u}] \rangle$ in (93c) relations such as (84) are needed (the right-amputated parts restored, however). Another typical bracket is $\langle [R_{z,x}, R_{y,u}] \rangle$, as in (95c). Using the known one-particle structure of the two factors, one derives (see Appendix A of Ref. 9)

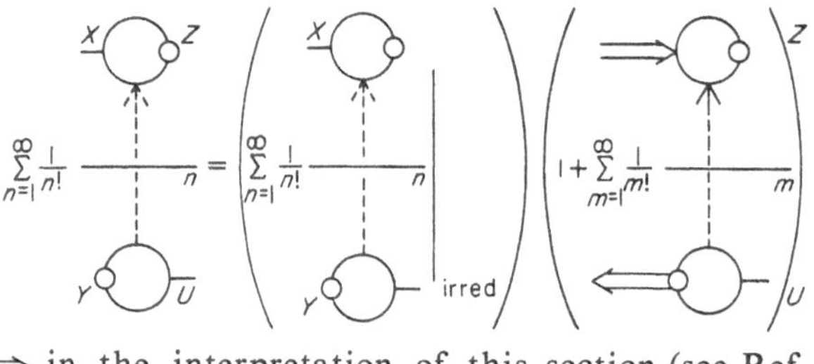

(97)

with $\Longrightarrow$ in the interpretation of this section (see Ref. 9). (The analogous integral equation for τ and $\bar{\tau}$ functions, without external sources, has been used in a certain "ladder" approximation by Fubini and coworkers for their peripheral model of high-energy scattering.)

For completeness, I will mention also that another system of nonlinear integral equations for G, such as suggested by the desire to raise the mass threshold that appears in the derivation of momentum-transfer analyticity (Ref. 26), with the aim of obtaining a larger analyticity region than usual, has been analyzed with similar results as in the case described before.

10. TRANSITION FROM r- TO τ-EQUATIONS

Equation (82) gives the prescription how to obtain from generalized functions by linear combination $\bar{\tau}$- and τ-functions and, generally, any vacuum expectation value $\langle \bar{T}(u_1 \cdots u_k) \cdot T(v_1 \cdots v_l) \rangle$. If we wish to obtain, e.g., the Bethe–Salpeter equation as simplest example, we need the four-point function with all arguments taking as well the retarded as the advanced role.

Again a matrix notation is advisable. Setting

$$f = \begin{bmatrix} \text{(diagram)} \end{bmatrix} \quad g = \begin{bmatrix} \text{(diagram)} \end{bmatrix} \quad h = \begin{bmatrix} \text{(diagram)} \end{bmatrix} = \sigma, g \quad j = \begin{bmatrix} \text{(diagram)} \end{bmatrix} = \sigma, f$$

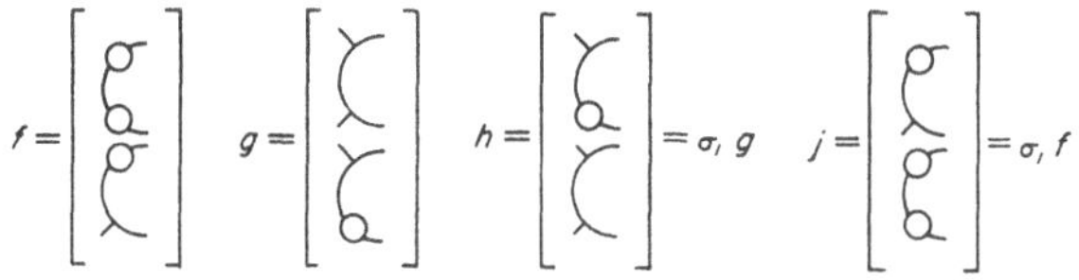

the 4×4 matrix

$$\mathscr{F} = \begin{pmatrix} f; g^T & f; j^T \\ h; g^T = 0 & h; j^T \end{pmatrix} = \begin{bmatrix} f \\ h \end{bmatrix} ; [g^T \ j^T]$$

$$= \begin{bmatrix} f \\ h \end{bmatrix} ; [f^T \ h^T] \, \Sigma_1 \, \sigma_1$$

$$\sigma_1 = \begin{pmatrix} \sigma_1 & 0 \\ 0 & \sigma_1 \end{pmatrix}, \quad \Sigma_1 = \begin{pmatrix} 0 & 1 \\ 1 & 0 \end{pmatrix} \tag{98}$$

contains all required functions. (We retain the convention $x^0 > y^0$, $z^0 > u^0$.)

We know already the ("retarded") Bethe–Salpeter equation for $\mathscr{F}_{11} = F$ and $\mathscr{F}_{22} = \tilde{F}$ from Section 9, while $\mathscr{F}_{21} = 0$. $\mathscr{F}_{12}$ can be similarly analyzed. It turns out that

$$\mathscr{F} = \mathscr{F}_i(1 + \mathscr{E}\mathscr{F}) \tag{99}$$

with $\mathscr{E}$ a triangular "unit" matrix, which is related to $\mathscr{F}$ as E is to F in Section 9. The triangular form of all matrices permits us to obtain $\mathscr{F}_{12}^i$ from the already determined $\mathscr{F}_{11}^i$ and $\mathscr{F}_{22}^i$. $\mathscr{F}_{12}$ can be expressed nonlinearly in an analogous manner as $\mathscr{F}_{11} - \mathscr{F}_{22}$ in (93a–d), and the analogous analysis that gives a decomposition of $\mathscr{F}_{12}$ in irreducible bracket can be carried out, such that

$$\mathscr{F} - \Sigma_1 \mathscr{F} \Sigma_1 \equiv \mathscr{D} \tag{100}$$

becomes a certain simple combination of commutators and anti-commutators.

One next observes that in (99) the intermediate $\mathscr{E}$ actually establishes all possible connections between the two points to the left and the two to the right, since only one connection that contradicts the time restriction, namely ⨉ , does not occur, with ⟺ lines between)— —(and ⟸ or ⟹ lines between)— ⟨ and ⟩ —(, respectively. Therefore, the time restriction does not destroy the covariance of the intermediate integrations in (99) similarly as it did not in (85).

Finally, we observe that, due to (98), symmetric $\mathscr{F}_s$ and $\mathscr{E}_s$ may be introduced with the help of $\sigma_1 \sum_1$

$$\mathscr{F}_s = \mathscr{F}_{is}(1 + \sigma_1 \sum_1 \mathscr{E}_s \sigma_1 \sum_1 \mathscr{F}_s) \equiv \mathscr{F} \sigma_1 \sum_1 \qquad (101)$$

Equation (83) suggests to regard the four-dimensional space of an index of $\mathscr{F}$, as a 2×2 direct-product space $1 \rightarrow 1 \cdot 1$, $2 \rightarrow 2 \cdot 1$, $3 \rightarrow 1 \cdot 2$, $4 \rightarrow 2 \cdot 2$ and to let the vector $a^T = [1, -(i/2)]$ be associated with the first, the vector $A^T = [1, -(i/2)]$ with the second index in the pair. We have, with the matrices introduced after (98)

$$\mathbf{a}; \mathbf{a}^T - \bar{\mathbf{a}}; \bar{\mathbf{a}}^T = -i\sigma_1,$$
$$\mathbf{A}; \mathbf{A}^T - \bar{\mathbf{A}}; \bar{\mathbf{A}}^T = -i\sum_1 \qquad (102)$$

Isolation in equation (101) of the $\tau(xyzu)$ function requires us to consider the matrix element $\mathbf{a}^T \mathbf{A}^T \mathscr{F}_s \mathbf{a} \mathbf{A}$. In fact, according to (82) application of $\mathbf{A}^T$ or $\mathbf{A}$ transforms the x, respectively, z argument, and of $\mathbf{a}^T$ or $\mathbf{a}$ the y, respectively, u argument, into a $-iT_x$, etc. Application of $\mathbf{A}^T$ etc., transforms into iT_x^+ etc. If in (101) (102) is now inserted, one obtains, if the external source j is not switched off, a "generalized" Bethe–Salpeter equation with 16 terms; if $j = 0$, only one term, the expected one containing only τ, τ^i, and the connecting Δ_F'-lines, survives, under the assumption that the linear combination of r-functions that gives $\bar{\tau}$, τ, etc., functions, which have the property that [cf. (82)] the sum of frequencies of the time-ordered arguments is nonpositive while that of the anti-time ordered arguments is nonnegative, leads for the r_i-functions to τ_i, $\bar{\tau}_i$, etc., functions that have the same property. Then a typical vanishing

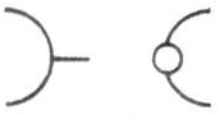

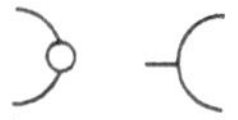

term in the BS-equation is

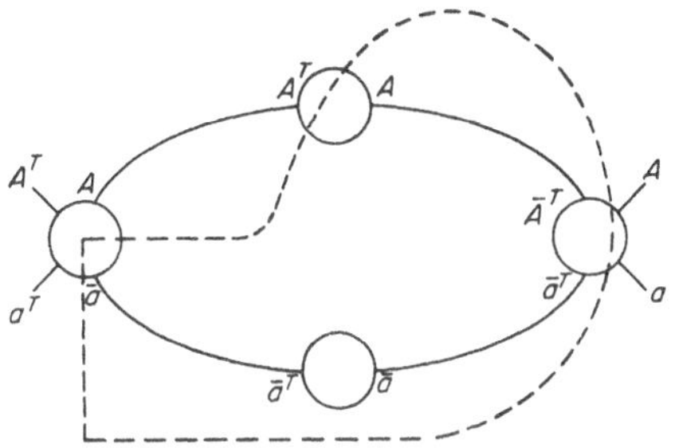

Here the sum of frequencies, fed into the part enclosed by the broken line, is positive (since only connected functions are considered), while it should be zero because of translation invariance. This is because the integrations actually are covariant ones as stressed before. That the frequency behavior of the irreducible functions is as assumed should follow from the study of the systems (93), (95), and others mentioned. A more direct way to convince oneself of this is the observation of Wraith[15] (who also has transcribed the method described here into a more flexible notation suitable for more general situations) that the frequency properties of r-functions reproduce themselves under composition such as in (101); the formal solution of (101) by

$$\mathscr{F}_{is} = \mathscr{F}_s - \mathscr{F}_s\sigma_1\Sigma_1\mathscr{E}_s\sigma_1\Sigma_1\mathscr{F}_s + \mathscr{F}_s\sigma_1\Sigma_1\mathscr{E}_s\sigma_1\Sigma_1\mathscr{F}_s\sigma_1 \times \Sigma_1\mathscr{E}_s\sigma_1\Sigma_1\mathscr{F}_s \mp \cdots$$

then gives the same frequency properties for $\mathscr{F}_{is}$ as for $\mathscr{F}_s$(if $j = 0$).

The Bethe–Salpeter kernel has here been obtained as a linear combination of two-particle irreducible retarded functions characterised by nonlinear systems such as (93), (95), and (100) with two-particle intermediate states in the $x, y|z, u$ channel removed, and by replacement of generally reducible by irreducible functions. The final step would be to translate these systems into irreducibility statements about τ and $\bar{\tau}$-functions alone. I did not do this but expect that the two-particle-irreducible forms of the systems corresponding to (31) or equivalently (30) and (32) for the special case under discussion are obtained.

It may be that this last result could also be obtained more directly, but it may not, in view of the much greater ease with which one can handle the retardedness property than the properties that characterize Feynman functions, as was stated earlier in connection with the perturbation theoretical verification of equation (50). Also, the fact that all derivations of analytic properties of scattering amplitudes etc. on the axiomatic level are based on retarded functions (and generalized retarded ones in the sense of Ref. 27, to which MPSA can be extended in a straightforward manner) makes the ordinary τ- and $\bar{\tau}$-functions appear as relatively uninteresting objects within the axiomatic approach.

11. CONCLUSION

As should have become clear from the results presented so far, there is no obstacle of principle (although that of rapidly increasing complication) for MPSA to be extended to more complicated situations, e.g., the removal of three-particle reducibilities etc. The method is always as follows:

I) Choose the Green's functions to be analyzed: Generalized r-functions (76); generalized r-functions in the sense of Ref. 27 and "mixed" generalized functions that are encountered in the analysis of the last-mentioned functions. External sources and functionals, thereof, may always be introduced by the replacement $A(x) \rightarrow R_x$;

II) Define irreducible functions (functionals) as solutions of Bethe–Salpeter-type equations chosen on the basis of perturbation theoretical expansions, and discuss the possible ("C.D.D")-sincularities of these functions;

III) Prove "linear" properties for these functions analogous to those that hold for the original functions;

IV) Prove the irreducibility (especially, absence of the intermediate states that were to be eliminated) of the absorptive parts of the irreducible functions (functionals).

Of these, step one is straightforward, step two less so due to the C.D.D.-singularities and perhaps need of subtractions as mentioned in Section 9, step three is again presumably straightforward (once two is satisfactorily solved) and step four, as experience with the two-particle case shows, is presumably quite complicated.

We mention briefly two extensions and one application of MPSA:

1) Eliminate low-lying intermediate states in several channels, e.g., in the case considered in Section 9 not only those in the $x, y|z, u$ channel but also in the $x, z|y, u$ and $x, u|z, y$ channels. The scheme to do this is quite simple in principle. For familiarity,

think of Feynman graph structures in considering the following equations (they apply, e.g., to a theory with the selection rule of pseudoscalar meson theory. Write them without that rule as an exercise!)

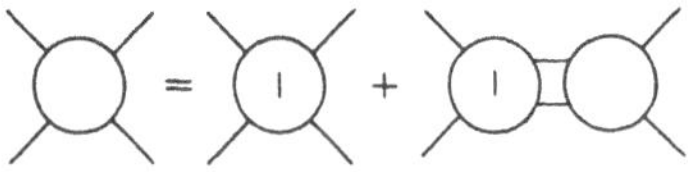

(103)

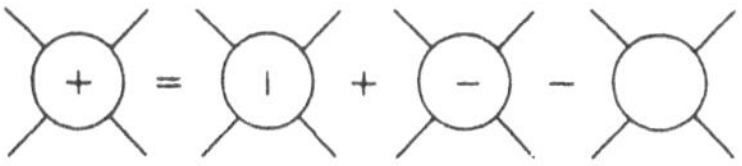

(104)

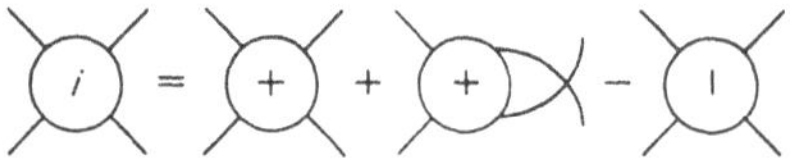

(105)

Here (103) is the Bethe–Salpeter equation for 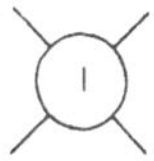 which is two-particle irreducible in the ← channel. Then (104) gives the function irreducible in the ↑ and ← channels, and (105) gives the function that is irreducible in all three channels. (Expressing the original function in terms of 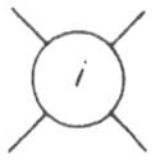 requires the solution of a nonlinear integral equation: Exercise!)

2) Consider also absorptive parts of an absorptive part (of absorptive parts . . .). Such an analysis would be necessary to obtain anything like Cutkosky's rules on an axiomatic basis. In the course of step IV for extension 1), absorptive parts of absorptive parts will in fact already come in.

3) Reference 10. If F^i in (86) or (88) has the properties implied by (93) and (95) (made irreducible) considered as linear conditions (i.e., with respect to support in momentum space, and that identical or related brackets be identical or so related), then the elastic scattering amplitude derived from G_{21} satisfies elastic unitarity up to the next mass threshold above the two-particle one in the $x, y| z, u$ channel. This implies, e.g., that ladder approximations and more generally all approximations to F_i by irreducible graphs lead to a scattering amplitude that satisfies unitarity as described. (The corresponding fact for Feynmam graphs may be seen elementarily). This "inversion theorem" clearly will have analogies in other situations.

I shall go back now to the statements I made in the first lecture, namely, that MPSA a) can be of help in the derivation of analytic properties of scattering amplitudes and other functions on the axiomatic basis, b) establishes the connection between axiomatic and Lagrangian field theory, and c) has an intuitive connection to "*S*-matrix theory."

a) The difficulty in the usual derivation of dispersion relations etc. are intermediate states of low mass. Zimmermann had shown[32] that in the simplest case, the π-N forward scattering dispersion relation, the convariant separation of the one-nucleon intermediate state (which invalidated the entirely elementary derivation of the dispersion relation by Goldberger[33]) was possible without losing retardedness. This was a simple case of one-particle-structure analysis and, in fact, the starting point of MPSA, which concerns itself with the generalization of this procedure. For example, for the $\langle N|\pi|N\rangle$ vertex, Jost[34] has shown that the "linear" properties of vacuum expectation values (cf. Schweber,[1] p. 741, 742 and pp. 776 ff.) do not suffice to derive a dispersion relation due to the small mass of the pion. "Unitarity" has to be used, e.g., it has to be exploited that continuum intermediate states of nucleon number one are not just those of, e.g., two particles of mass $(M_N + M_\pi)/2$ each, but that they contain at least one nucleon, and also that all particles in these states interact causally. For example note that in graphs

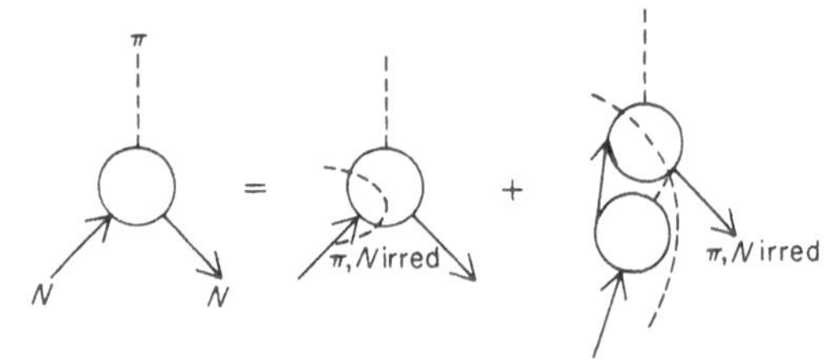

(106)

the last term can be further decomposed into

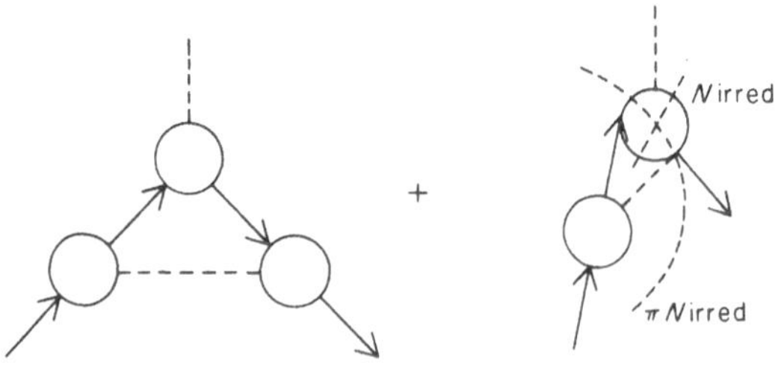

(107)

The first term on the right-hand side of (106) already possesses a raised mass threshold; the first term in (107) is comparatively simple but the second term is not. Comparison with perturbation theory suggests that all terms should satisfy the expected dispersion relation separately. It would be interesting to insert in the first term of equation (107) vertices that have already as many of the linear properties as one has been able to derive for the three-point function, and to carry out the integration in terms of integral representations. If the result gives an analyticity region larger than before, this would be an indication of the effectiveness of MPSA. However, the linear program is hereby not circumvented; not only is the first term on the right-hand side of (106) similar to the left-hand side but the second term in (107) is especially tough since it contains a four-point function. Thus, it seems that any usefulness of MPSA in the sense discussed here must await progress in mastering the "linear program."

b) The connection to Lagrangian quantum field theory[35] is very simple; it is stated in Ref. 9 and elaborated in connection with renormalisation in Ref. 12 that, e.g., in a theory with trilinear interaction, like quantum electrodynamics (QED) Green's functions have the property that those that are two-particle irreducible with respect to any single argument vanish except for the bare vertex. For example, in *Ps-Ps* meson theory the first term on the right-hand side of (106) would be a constant, which might be absorbed in the subtraction required for the second term in perturbation theory. However, that the first term is trivial is a strong statement. A most interesting but presumably rather difficult question to which I have no answer or even clue to offer now, is: What effect does the first term have on observables (e.g., scattering amplitudes) when one is not dealing with the most general axiomatic theory of the given stable particles, symmetries and conservation laws, but with a theory specified by a Lagrangian? For example, in QED with a theory not giving a Pauli-moment to the electron, assuming perturbation theory is not to be relied upon to rule it out.

c) The intuitive connection to *S*-matrix theory, the latter understood in the sense that perturbation theoretical analyticity should be preserved as far as possible, consists in the observation of Section 4 of Ref. 9 that carrying MPSA for a Green's function to higher and higher mass states gives a picture of the function, in a sense, essentially identical to the perturbation theoretical one with, however, bare vertices replaced by higher-irreducible functions that can be expected to behave relative to comparatively small momenta like point vertices. "First sheets" become important: infinite sums are not allowed, but integral equations must be solved instead. This is analogous to the need one has in S-matrix theory to obtain the possible singularities in higher sheets by analytic continuation from the "first sheet" on the basis of unitarity. One characteristic difference between MPSA and perturbation theoretical analyticity, in the sense of Landau, may be the possibility of C. D. D-singularities of irreducible functions in the former and their absence in the latter. As to such singularities, the following model might be instructive though not really relevant (cf. also the model of Section 7). Assume the four-point function to have in the $\uparrow$ channel an "unstable particle," e.g., to have (in terms of Lagrangian field theory rather than in the sense of Section 8) the decomposition

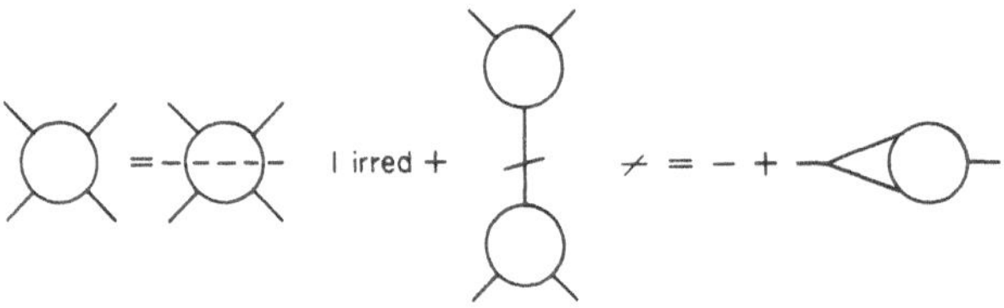

where

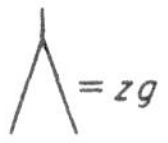

with

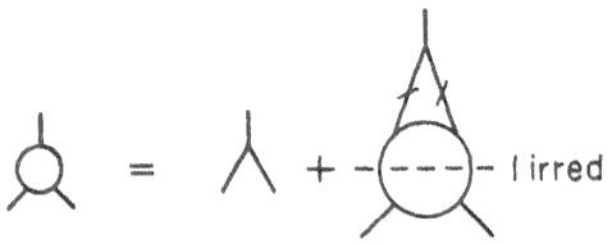

where however + does not have a stable-particle pole. Then, if

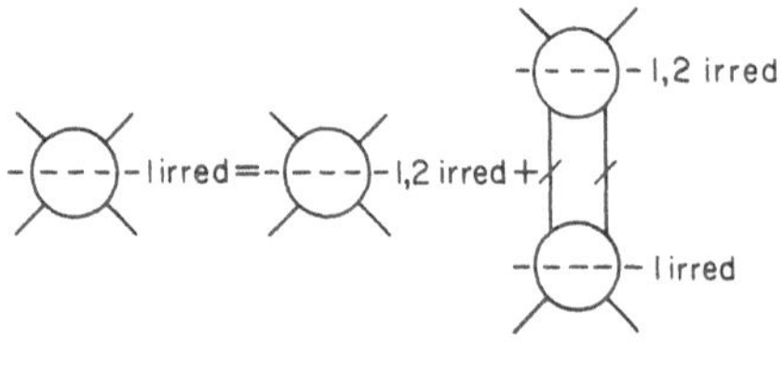

the kernel 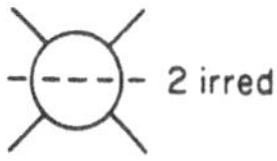

defined by

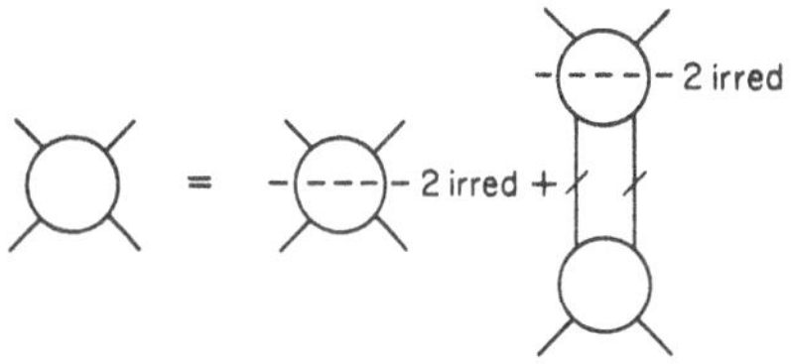

is

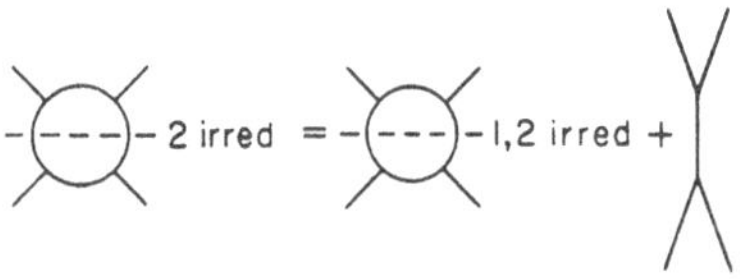

i.e., has a pole at the bare mass of the "unstable particle."

As to the actual achievement of MPSA at the present time, one may say that it shows that the natural setting of any Bethe–Salpeter-type structure of Green's functions is axiomatic field theory; this frame allows us to derive (on a level of rigor discussed in detail before) the irreducibility properties of the irreducible functions in precise form, and inversely to prove that unitarity is satisfied up to higher mass thresholds (provided those irreducibility properties are satisfied). The simple Bethe–Salpeter picture (think, e.g., of irreducible graphs containing stable-particle lines only) is slightly complicated by the C.D.D-singularities about which further analysis would be necessary. There is, however, no need (and, perhaps, even no room) for speculation.[35]

REFERENCES

1 R.F. Streater and A.S. Wightman, "PCT, Spin & Statistics, and all that," Benjamin, New York, 1964, S.S. Schweber, *Introduction to Relativistic Quantum Field Theory,* Row, Peterson and Co., Evanston, Illinois (1961).

2 D. Ruelle, *Helv. Phys. Acta* **35**: 147 (1962); cf. also: R. Jost, "The General Theory of Quantized Fields," American Mathematical Society, Providence, Rhode Island (1965).

3 K. Hepp, *Helv. Phys. Acta* **37**: 639 (1964).

4 J. Bros, H. Epstein, and V. Glaser, *Nuovo Cimento* **31**: 1265 (1964).

5 S. Mandelstam, *Nuovo Cimento* **15**: 658 (1960).

6 D. Ruelle, unpublished.
R.F. Streater, *Nuovo Cimento* **25**: 274 (1962).

[7] H.P. Stapp, Matscience Report 26.
[8] L.D. Landau, Proceedings of the Kiev Conference (1959).
[9] K. Symanzik, *J. Math. Phys.* **1**: 249 (1960).
[10] K. Symanzik, Proceedings of the Rochester Conference 1960, p. 211.
[11] K. Symanzik, in *Lectures in Theoretical Physics*, Boulder, 1960, p. 490, ed. W.E. Brittin *et. al.* Intersc. Publ. (1961).
[12] K. Symanzik, in *Lecutures on High-Energy Physics*, p. 485, ed. B. Jakšić Gorden and Breach, New York (1966).
[13] K. Symanzik, unpublished.
[14] G.C. Wraith, *Nuovo Cimento* **21**: 352 (1961).
[15] G.C. Wraith, *Nuovo Cimento* **27**: 561 (1963).
[16] W. Zimmermann, *Nuovo Cimento* **10**: 597 (1958).
[17] V. Volterra, *Theory of Functionals*, Dover Press (1959).
[18] H. Lehmann, K. Symanzik, and W. Zimmermann, *Nuovo Cimento* **1**: 209 (1955).
[19] H. Lehmann, K. Symanzik, and W. Zimmermann, *Nuovo Cimento* **6**: 319 (1957).
[20] V. Glaser, H. Lehmann, and W. Zimmermann, *Nuovo Cimento* **6**: 1122 (1957).
[21] F.J. Dyson, *Phys. Rev.* **82**: 428 (1951).
[22] H. Lehmann, *Nuovo Cimento* **11**: 342 (1954).
H. Umezawa and S. Kamefuchi, *Prog. Theor. Phys.* **6**: 543 (1951).
[23] L. Castillejo, R. H. Dalitz, and F. J. Dyson, *Phys. Rev.* **101**: 453 (1956).
[24] M. Veltman, *Physica* **29**: 186 (1963).
[25] H. J. Bremermann, R. Oehme, and J. G. Taylor, *Phys. Rev.* **109**: 2178 (1958).
[26] H. Lehmann, *Suppl. Nuovo Cimento* **14**: 153 (1959).
[27] O. Steinmann *Helv. Phys. Acta* **33**: 257, 347 (1960).
D. Ruelle, *Nuovo Cimento* **19**: 356 (1961).
A. Araki, *J. Math. Phys.* **2**: 163 (1961).
[28] R. Haag, *Phys. Rev.* **112**: 669 (1958).
[29] W. Zimmermann, *Nuovo Cimento* **15**: (1960).
[30] B. Ferretti, *Nuovo Cimento* (L) **12**: 457 (1954).
[31] A. Salam, *Phys. Rev.* **82**: 217 (1951).
[32] W. Zimmermann, *Nuovo Cimento* **13**: 503 (1959).
[33] M. L. Goldberger, *Phys. Rev.* **97**: 508 (1955).
[34] R. Jost, *Helv. Phys. Acta* **31**: 263 (1958).
[35] Cf. also: J. G. Taylor, *Suppl. al Nuovo Cimento* **1**: 857 (1963).

Author Index

A

Adams, J. B., 61
Ademollo, M., 36
Araki, A., 143, 170
Argo, H. V., 56
Aubert, B., 35

B

Bardacki, K., 36
Barrett, B., 36
Bartlett, M. S., 108, 120
Barut, A. O., 97
Behr, A., 35
Belinfante, 78, 80
Bernardini, G., 35
Bernstein, J., 37
Bhabha, H. J., 97
Bialynicki, I., 64, 74
Birula, 64, 74
Bleser, E., 45, 61
Block, B. L., 56, 62
Blockimtsev, L., 58
Block, M., 35
Bogoliubov, 122
Bouchiat, C., 36
Breman, J. B., 56, 62
Bremermann, H. J., 170
Brene, N., 36
Bros, J., 122, 169
Brown, C., 119, 120
Burgman, J. O., 56, 62
Burnstein, A., 36

C

Cabibbo, N., 36
Campbell, N., 113, 120
Christenson, J. H., 36
Clay, D., 62
Cohen, R. C., 62
Collard, H., 62
Conforto, G., 43, 61
Cornwall, J. M., 36
Courant, H., 36
Cronin, J. W., 36

D

Dalitz, R. H., 36, 170
Day, T. B., 36
Delbourgo, R., 36
Delorme, J., 59
d'Espagnat, B., 36
de Swart, J. J., 36
Dirac, P. A. M., 75, 78, 80, 97
Drechster, W., 57
Duck, I., 60, 62

E

Edelstein, R. M., 57
Epstein, C. H., 122, 169
Ericson, T., 59, 62

F

Falomkin, I. V., 57
Ferretti, B., 157, 170
Fetkovich, J. G., 56, 62
Feynman, R. P., 35

Fields, T. H., 62
Fierz, 78, 80
Filippov, A. I., 62
Filthuth, H., 36
Fischer, J., 62
Fitch, U. L., 36
Flamand, G., 56
Focardi, S., 61
Foldy, L. L., 62
Ford, K. W., 56, 62
Freund, P. B., 36
Fried, H. M., 64, 74
Fubini, S., 37
Furlan, G., 36

G
Gatto, R., 36
Gelfand, M., 97
Gell-Mann, M., 35, 37
Gershtein, S. S., 61
Glaser, V., 122, 169, 170
Glashow, S., 35
Glasser, R. G., 36
Goldberger, M. L., 166, 170
Gross, L., 97
Gursey, F., 36

H
Haag, R., 123, 150, 170
Halpern, A., 62
Harish-Chandra, 78, 85, 97
Harrison, F. B., 62
Hellesen, B., 36
Hepp, K., 122, 169
Herz, A. J., 36
Hill, R. E., 62
Hull, A. W., 120

I
Izuyama, T., 61

J
Jenkino, D. A., 62
Johnson, K., 69, 113, 120
Jost, R., 168, 169
Judd, D. L., 61

K
Kamefuchi, S., 170
Kehoe, B., 36
Keuffel, J. W., 62
Khalatnikov, J. M., 64, 74
Kikuchi, R., 120
Kirsch, L., 36
Kruse, H. W., 62
Kulyukin, M. M., 62
Kummer, W., 37

L
Lach, J. T., 62
Lai, C. S., 37
Landau, L. D., 64, 74, 122, 124, 170
Lederman, L., 61
Lee, B. W., 36
Lehmann, H., 122, 143, 170
Leontic, B., 62
Levy, M., 36
Lipman, N. H., 62
Lovelace, C., 36
Love, W. A., 56, 62
Lowys, J. P., 35
Lundby, A., 62

M
Maier, E. J., 56, 62
Mandelstam, S., 122, 169
Marder, S., 62
Marshak, R. E., 35
Mathews, P. M., 120
McGuire, A. D., 62
McIlwain, R. L., 62
Meunier, R., 62
Meyer, P., 36
Migdal, A. B., 99, 100
Mills, R., 70, 74
Minguzzi-Ranzi, A., 36
Minlos, R. A., 97
Mittner, P., 35
Mizuno, Y., 61
Morita, M., 61, 62
Moullin, E. B., 114, 120
Moyal, J. E., 120

N
Nadelhaft, I., 62
Naimark, M. A., 74
Naunenberg, M., 36
Ne'eman, Y., 35

O
Oakes, R. J., 36
Okubo, S., 36, 64, 74
Ortin-Lecourtois, A., 35

P
Pais, A., 36
Pauli, 78, 80
Plano, R. J., 36
Pontecorvo, B., 62
Prentki, J., 36
Primakoff, H., 56, 57

R
Radicati, L., 36
Ramakrishnan, A., 120
Rarita, 78
Reynolds, G. T., 62
Riazuddin, 36
Riddell, R. J., 61
Roman, P., 97
Roos, M., 36
Rosen, J., 61
Rothberg, J., 61
Rowland, 112, 114, 120
Ruelle, D., 143, 169

S
Sakita, B., 36
Salam, A., 35, 157
Scarl, D. B., 62
Scherbokov, Yu. A., 62
Schweber, S. S., 121, 166, 169
Schwinger, J., 35, 74, 78
Sechi-Zorn, S., 36
Seeman, N., 36
Segar, A., 36
Segal, I. E., 97
Segré, G., 37, 58, 62
Sens, J. H., 59
Shapiro, J., 58, 62
Shapiro, Z. Ya., 97
Shimzu, M., 61
Siegel, R. T., 62
Stapp, H. P., 170
Snow, G. A., 36
Srinivasan, S. K., 170
Stech, B., 57, 62
Steinberger, J., 36
Steinmann, O., 143, 170
Strathdee, J., 36
Streater, R. F., 121, 124, 151, 169
Stroot, J. P., 62
Sudarshan, E. C. G., 35, 120

T
Taylor, A. E., 62
Taylor, J. G., 170
Teja, J. D., 62
Treiman, S. B., 36
Thirring, W., 37
Truong, T. N., 36
Tsuppo-Sitinkov, V. M., 62
Turlay, R., 36
Twiss, R. Q., 120

U
Uberall, H., 59, 62
Umezawa, H., 97, 170

V
Vasudevan, R., 115, 120
Veltman, M., 139, 170
Volterra, V., 170

W
Ward, J. C., 36
Weinberg, S., 36, 47, 61, 97
Werntz, C., 57, 62
Wightman, A. S., 121, 169
Wigner, E. P., 81, 82, 97
Williams, N. H., 113, 120
Willis, W., 24, 36
Wolf, E., 120
Wolfenstein, L., 56, 62
Wraith, G. C., 124, 163, 170

Y
Yamaguchi, Y., 57
Yang, C. N., 70, 74
Yennie, D. R., 68, 74

Z
Zaimidoronga, O. A., 62
Zavattini, E., 61
Zdanis, 62
Zeldevich, Ya. B., 61
Zimmermann, W., 124, 151, 170
Zumino, B., 64, 69, 74
Zwanziger, D., 95, 97
Zweig, G., 35

Subject Index

A
A.c.a.c. hypothesis, 42, 49, 61
A^4-coupling, 157
Adams' formula, 43, 47
Advanced singularities, 143
Age-dependent birth and death
processes, 108
Algebra of $U(4)$, 5
Amplification factor, 113
Amplitude renormalization, 136
Amputated function, 140, 145
Amputation, 139, 148
Analytic properties, 122
Analytic S-matrix theory, 123
Anode circuit, voltage fluctuations in, 112
Anomalous magnetic moments, 75
Arbitrary vector currents, 63
Asymptotic particle states, 121, 135
Atomic capture rate, 39, 41, 46
in hydrogen, 47
Axial
current, 7
vector coupling constant, 41
vector currents, 2, 26, 29, 42
Axiomatic quantum field theory, 121, 166

B
Barkhausen noise, 107, 115, 116, 165
Baryon matrix element, 26
Basic triplet fields, 28
Bethe-Salpeter
equation, 99, 124, 157, 161, 162, 163
picture, 169
techniques, 100
Bhabha's theory, 85
Binary mixtures, 119
Bohr radius, 54
Bose statistics, 155

C
Cabibbo
angle, 1, 20
model, 2
Canonical commutation relation, 136
Capture rate in He^3, 48
Casimir operator, 82
C.D.D. singularities, 145, 149, 164, 168
C.D.D. zeros, 140
Charged currents, 4
Clebsch – Gordan coefficients, 17
Closed algebra, 5
Commutation relations 2, 11, 29, 31, 33
Compact group, 5
Completeness axiom, 121, 122
"Composite" particles, 124
Condensed graphical notation, 139

Conditional probability frequency functions, 109
Conformal group, 83
Conjugation parity, 40
Conservation laws, 104
Connected r-functions, 153
Connected τ-functions, 153
Conserved vector current, 22
Correlation functions, 72, 111, 118, 119
Counter term, 136
Coupling between vector and pseudoscalars meson, 27
CP-violating interactions, 6
Cumulants in many-body problems, 153
Current—current coupling, 6
Cutkosky's rules, 165
C.v.c. hypothesis, 60

D
Damping factor, 31
Definite G-parity, 60
Density of arrivals, 114, 115
D/f ratio, 22, 23
Dipole
  γ-absorption, 100
  response fucntions, 100, 101
  resonance, 105
  term, 101, 104
Dirac-Fierz-Pauli theory, 85
Dirac
  equation, 79
  spinors, 79
Dispersion relation, 143
Dotted spinors, 76
Doubled graphs, 137, 138, 144, 152
Dyson's doubled graphs, 137, 138, 144, 152
Dyson expansion, 87

E
Edge of wedge theorem, 143
Effective charges, 53, 105
Effective internal resistance, 114
Elastic γ-ray, scattering, 100
Electromagnetic
  corrections, 22
  current, 8, 11, 12
  interactions, 4
Electron propagator, 68
Elementary particles, 124
Elliot-Flower wave function, 55
Energy denominators, 28
Expectation values, 101
External current, 67
External fields, 66

F
Fermi liquids, theory of, 99
Feynman
  amplitude, 132
  boundary condition, 139
  disentangling method, 152
  graphs, 124, 136, 137, 138, 144, 165
  integral, 67
  rules, 87, 91, 93
Forward light cone, 152
Fredholm method, 157
f-type matrix, 18
Functionals, 126
Functional derivatives, 127
Fundamental triplet, 12

G
Galilean invariance, 100, 104, 105
Gauge invariance, 65, 100, 104
Gelfand construction, 64
General relativity, 123
Generalized retarded functions, 143, 154
Generalized r-functions, 153, 154
Generating functionals, 65, 129
Goldberg—Treiman relations, 23 24, 26, 28, 42
Graphical notation, 154
Green's functions, 121, 164, 167
G-spin, 15

H
Haag–Araki approach, 122
Haag–Ruelle scattering theory, 122
Hadron currents, 1, 7, 14, 16, 25
Hartree–Fock approximation, 99, 101
Heisenberg representation, 130
Hermitian scalar local field, 124
Higher-order corrections, 102
Hilbert space, 130
Hook's law, 128
Hypercharge, 11
Hyperon decays, 23

I
Impulse approximation, 52, 61
Induced pseudoscalar coupling constant, 55
Induced pseudoscalar term, 23
Inelastic scattering, 100
Infinite-dimensional quantum gauge transformation, 64
Infinitesimal gauge transformation, 70
Influence functional, 129
Inhomogeneous Poisson processes, 108, 109
Inhomogeneous proper Lorentz group, 121
Integrability condition, 128
Interaction Lagrangian, 12
Intermediate nuclear model, 56
Invariance, 121
Invariant functions, 133, 134
Irving wave function, 57
Irreducible doubled graphs, 144
Irreducible functional, 155
Isopin current, 6
I-spin current, 11, 13

K
K_{e3}-decay, 20
K-electron capture, 39
K-field, 30
Kinetic terms, 6
Klein–Gordon operator, 141, 149
  equation, 83, 125, 130
K-mesons, 28
K-mesons dominance model, 33
Kolmogorov equations, 109, 112

L
Ladder approximation, 160, 165
Lagrangian quantum field theory, 122, 166
Landau gauge, 65
Lehman weight functions, 139
Lepton–baryon symmetries, 13
Lepton current, 12, 51
Lepton–lepton coupling, 1
Lepton-number current, 4
Lepton operator, 53
Leptonic decays of hyperons, 20
Leptonic transitions, 24
Leptonic weak interaction, 1
Leptons, 7
Lie gauge groups, 70
Liquid hydrogen, 39, 43
Little group, 82, 83
Local operator rings, 122
Longitudinal part, 64
Lorentz
  covariant densities, 10
  gauge, 70
  group, 76
  group in five fimensions, 87
  invariance, 60
LSZ approach, 122, 123
LSZ asymptotic condition, 131
Lubanski invariant, 82

M
Macroscopic magnetization, 117
Magnetic coupling constant, 41
Magnetic form factors, 22
Mandelstam representation, 122
Many-particle
  intermediate state, 144
  structure of analysis, 121
  structure of Green's functions 121

Markovian processes, 107
Mass
  renormalization, 136
  shell, 135
  spectrum conditions, 121
  threshold, 148
Matrix-algebraic approach to relativistic wave equations, 84
Mean-square response, 115
M-functions, 96
Microscopic magnets, 117
Microscopic Weiss fields, 117
Mixed commutation, 32
Molecular capture rate, 39, 43
Momentum transfer, 122
Muon capture, 39

N
Ne-deuterium concentration, 43
Neutron β-decay, 15
Neutral lepton currents, 4
Non-Abelian gauge fields, 70
Nonchanging currents, 1
Nonleptonic interactions, 2
Nonlinear field theory, 137
Non-Markovian process, 107, 108, 109, 112, 116, 118
Nonperturbation theoretical approach, 124
N-point propagator, 68
Nuclear β-decay, 22
Nucleon–nucleon scattering, 122
Nucleon operator, 53
Null spinors, 95

O
Octet of vector currents, 20
One-particle
  cuts, 144
  intermediate states, 144, 148
  irreducible functions, 148
  poles, 140
Orthonormal set, 130
Overlapping pulse, 107

P
Partial muon capture rate, 49, 53
Pauli spin operator, 41
Pauli spinors, 41
Perturbation theory, 123, 124
Poincaré group, 82
Poisson parameter, 119
Positive-frequency solutions, 130
Probability frequency functions, 109
Product densities, 118
Propagator, 92
Pseudoscalar and scalar densities, 22, 29
Pseudoscalar couple constant, 24

Q
Quark model, 16
Quasi-particle description, 99

R
Raleigh expansion, 54
Random-phase approximation, 101
Random times, 107, 117
Rarita-Schwinger
  equation, 81
  theory, 80
Reducible doubled graphs, 144
Reduction formula, 132
Regge poles, 75, 157
Relativistic form factors, 57
Relativistic wave equation for arbitrary spin, 78
Renormalization, 2
Renormalization corrections, 12, 25
Response function, 100, 119
Retarded multiple commutators, 133
Retarded singularities, 143
r-functions, 131, 133, 134, 136, 137, 139, 153, 154
Rowland theory, 114
R-products, 133

S
Scalar representation, 83
Scalars, 29
Scattering amplitude, 122, 123
Second-rank spinor, 77
Selection rules for muon capture in nuclei, 55
Semi-leptonic, 1
Semi-leptonic transitions, 12
Shapiro ratio, 57
Shell model, 101
Shot effect, 114
Shot noise, 107, 108, 112
Single-particle dressing, 101
Single term, 21
Singularities, 135, 143, 144
Singularity of a τ-function, 132
S-Matrix theory, 167
Spacelike surface, 125
Spectral decomposition, 104
Spinor
  derivatives, 77
  electrodynamics, 65
  representation, 76
Spin-tensors, 78
Stationary system, 113
Statistical independence, 107
Stochastic
  point process, 111
  processes, 107
Strangeness
  changing axial vector amplitudes, 26
  changing currents, 1, 11, 35
  changing density, 30
  changing parts, 14
  changing transition, 19, 25
  non-changing currents, 1
  non-changing parts, 14, 15
$SU(3)$-symmetry, 21, 25, 27, 32, 34
Symmetry breaking, 25

T
T and R products, 131
τ-functions, 131, 132, 133, 136, 139, 153, 154
Tensor representation, 76
Three-dimensional rotational group, 76
Three-particle states, 148
Threshold singularities, 144
Time reversal invariance, 60
T-matrix, 99
Transition rates, experimental, 43
Transformation to quasi-particles, 103
"Trees", 152
Triplet number, 10, 11, 14, 21
Two-particle
  C.D.D. singularities, 160
  intermediate states, 148
  structure, 154
Two-point photon propagator, 64

U
$U(2)$-algebras, 14
$U(3)$-matrices, 9
Unamputated functional, 155
Unitarity, 123
Unitary scattering, 131
Universal
  axial vector damping factor, 27
  damping, 25, 26
  suppresion, 2
  $V-A$ theory, 40
  weak coupling, 12
  weak interactions, 1
Universality
  hypothesis, 22
  principles, 15
Unrenormalized quantities, 135

V
Vacuum expectation value, 133
Vacuum graphs, 136
Vector and axial vector currents, 18

Vector coupling constant, 41
Vector meson baryon coupling, 27
Vertex functions, 69
Vertex renormalization, 101
Voltage fluctuations, 112
Volterra expansion, 64, 128, 136
  series, 127
V-spin currents, 11, 13

W

Wave renormalization constant, 69
Weyl equation, 79
Wick product, 137
Wick's rule, 91
  theorem, 130
Wightman approach, 122
Wigner–Eckhart theorem, 17